Time Map Phonology

Text, Speech and Language Technology

VOLUME 5

The titles published in this series are listed at the end of this volume.

Time Map Phonology

Finite State Models and Event Logics in Speech Recognition

by

Julie Carson-Berndsen
University of Bielefeld,
Faculty of Linguistics and Literary Studies

KLUWER ACADEMIC PUBLISHERS
DORDRECHT / BOSTON / LONDON

A C.I.P. Catalogue record for this book is available from the Library of Congress.

ISBN 0-7923-4883-4

Published by Kluwer Academic Publishers,
P.O. Box 17, 3300 AA Dordrecht, The Netherlands.

Sold and distributed in the U.S.A. and Canada
by Kluwer Academic Publishers,
101 Philip Drive, Norwell, MA 02061, U.S.A.

In all other countries, sold and distributed
by Kluwer Academic Publishers,
P.O. Box 322, 3300 AH Dordrecht, The Netherlands.

Printed on acid-free paper

Printed in the Netherlands.

To
Ron Carson

CONTENTS

PREFACE

This book is a revised version of my doctoral thesis which was submitted in April 1993. The main extension is a chapter on evaluation of the system described in Chapter 8 as this is clearly an issue which was not treated in the original version. This required the collection of data, the development of a concept for diagnostic evaluation of linguistic word recognition systems and, of course, the actual evaluation of the system itself. The revisions made primarily concern the presentation of the latest version of the SILPA system described in an additional Subsection 8.3, the development environment for SILPA in Section 8.4, the diagnostic evaluation of the system as an additional Chapter 9. Some updates are included in the discussion of phonology and computation in Chapter 2 and finite state techniques in computational phonology in Chapter 3.

The thesis was designed primarily as a contribution to the area of computational phonology. However, it addresses issues which are relevant within the disciplines of general linguistics, computational linguistics and, in particular, speech technology, in providing a detailed declarative, computationally interpreted linguistic model for application in spoken language processing. *Time Map Phonology* is a novel, constraint-based approach based on a two-stage temporal interpretation of phonological categories as events. It takes recent developments in multi-linear phonological description into account but goes further than previous formal interpretations of autosegmental phonology in that phonological relations are explicitly related to intervals in actual speech signals as required by a spoken language recognition system. The constraint-based event model of *Time Map Phonology* has been provided with a procedural interpretation and has been operationalised by implementation and testing within the linguistic word modelling component of a spoken language recognition system.

There are two technical details in connection with the text which should be mentioned in advance. Throughout the text of the book, the IPA is used for transcriptions; in the figures, the SAMPA phonetic alphabet (Wells, 1989) is used. The SAMPA phonetic alphabet for German is provided in Appendix A. Familiarity with the elements of formal languages such as the Chomsky hierarchy and automata theory is assumed in this work. A brief definition of the Chomsky hierarchy is provided in Appendix B.

ACKNOWLEDGEMENTS

I would like to thank, in particular, Dafydd Gibbon for excellent supervision and direction, for helpful comments on earlier drafts of the thesis and for keeping me to the task. Thanks go also to my colleagues Doris Bleiching, Gunter Braun, Hagen Langer and Martina Pampel for fruitful discussions in connection with various aspects of the research for this work. Particular thanks go also to an anonymous reviewer who provided detailed comments and criticisms on the text. However, I am solely responsible for any errors or inconsistencies which may remain.

Special thanks go to my family, in Ireland and Germany, for their support and encouragement over the years. This book is dedicated to the memory of my brother, Ron Carson, who remained my example throughout.

1

INTRODUCTION

This book presents a novel, constraint-based approach to phonological parsing based on temporal interpretation of phonological categories as events and combining aspects of event logic with efficient processing strategies. *Time Map Phonology* takes recent developments in autosegmental phonological description into account but goes further than previous formal interpretations of autosegmental phonology (cf. Bird & Klein, 1990; Kornai, 1991) in that phonological relations are explicitly related to intervals in actual speech signals as required by a spoken language recognition system. The constraint-based event model of *Time Map Phonology* has been provided with a operational interpretation and has been implemented and tested within the linguistic word modelling component (BELLEx3, cf. Gibbon et al. 1992) of a spoken language recognition system.

The aim of computational phonology is to design computationally interpretable declarative models of phonological description which can be used for both generation and analysis purposes. This would seem to be an interesting challenge for spoken language recognition where declarative phonology could play a major role. Phonological parsing is the analysis of phonetic information on the basis of phonological knowledge about a language; providing phonetic input with a structure (often in terms of syllables) which is relevant at the phonological level. However, despite a number of developments in connection with phonological parsing (e.g. Church, 1983, 1987b; Carson, 1988, Gibbon & Braun, 1988a; Carson-Berndsen et al. 1989), phonological theory is only now beginning to find its place in spoken language recognition (cf. the model proposed in Kirchhoff, 1995b), although current successful approaches to speech recognition do implicitly avail of elementary phonological structure (e.g. Hidden Markov models, demisyllable-based recognition). Developments in the area of phonological parsing and the reasons why they are only now being considered by the speech community are discussed below.

1.1 SPOKEN LANGUAGE RECOGNITION AND PHONOLOGICAL PARSING

It would go beyond the bounds of this chapter to provide yet another overview of current speech recognition systems. Recent overviews can be found in Sagerer (1990), Kunzmann (1991) and more generally, in terms of the methods used for speech recognition, in Holmes (1988). The collection of papers in Niemann et al. (1988) and the volume edited by Waibel & Lee (1990) provide an excellent overview of current trends in speech recognition and notably not one of these papers addresses the issue of phonological parsing. Rather phonological parsing has been a topic which has interested computational linguists, psychologists and, to a surprisingly limited extent, phonologists (but cf. Kaye, 1989). In this section, the main direction of spoken language recognition is discussed briefly and the notion of phonological parsing is introduced.

Most current models of speech recognition make a distinction between a word recognition or *speech component* and a linguistic or *language component*. The *language component* usually refers to processing above the level of the word while processing below the word level is assumed to be a search task of the *speech component*. A general distinction can be made in connection with the *speech component* of a spoken language recognition system between data-driven (or speech engineering methods) and knowledge-based approaches (Holmes, 1988). The former refer to more mathematical methods, stochastic or pattern recognition methods the most common of which are *Dynamic Time Warping* (DTW) and *Hidden Markov Models* (HMMs). The dynamic time warping approach searches for an optimum match between the feature vectors of a stored word-template and a signal token using a non-linear temporal alignment path. Phonological or linguistic information does not play a role. Hidden Markov models are similar to probabilistic finite state networks and are in a sense a further development of DTW in that they model temporal and spectral variability. In contrast to traditional finite state automata, the choice of states and transitions is governed by a probability distribution (cf. Rabiner, 1989). A hidden Markov model generates patterns of feature vectors to represent a word to be recognised and the probability distributions provide the probability that a particular feature vector will be generated when the model is in a particular state. Since the model is based on the interaction between two sets of probabilities, the knowledge structure (i.e. the state sequence) is *hidden* in the model and is not explicit as in finite state automaton representations of phonological knowledge. Statistical training can lead to the compilation of different hidden Markov models and the knowledge represented by the state sequence is elementary and incomplete in the linguistic sense. However, if the structure for the Markov model is provided *a priori* and the probabilities calculated on this basis, the model is closer to a probabilistic version of a phonotactic finite state machine. It is important to note that in linguistic terms and contrary to common formulations in speech technology a HMM encodes *phonological* and not *phonetic* information. The latter is encoded in the feature vectors.

Knowledge-based methods, on the other hand, refer to artificial intelligence techniques in terms of expert systems with production rules. This is a more linguistic approach which is, however, seldom found in current speech recognition systems. While there would seem to be a moral to this, namely that the data-driven approaches are more successful and are therefore more widely applied, this book aims to show how phonological processing can be included more successfully into a spoken language recognition system by addressing fundamental issues of linguistic competence before tackling performance issues of various kinds. Holmes (1988) makes the following claim in this connection:

> At present the *data-driven* statistical methods seem to give better performance than *knowledge-based* methods on most speech recognition tasks, although we know that as presently formulated they ignore much of the information in the acoustic signal that we know is important in human speech recognition. There is thus a strong implication that we ought to advance the state of the art by combining the best aspects of both methods. In principle the ability to learn by example that is characteristic of the data-driven approach could be extended to learn much more complicated phonetic features. All that is needed is to give the models a rich enough structure and then present a sufficient quantity of labelled training data. I believe that this method is not likely to work in practice for any feasible type of model and amount of training, unless we use our knowledge of the particular types of feature that are phonetically important when choosing the structure of the models. We could thus design models to cope only with the sort of features that our knowledge leads us to believe might be cues for phonetic distinctions. If model structures of these types can be specified first, using human knowledge, it will obviate the need for the learning process to discover the types of feature, in addition to their statistics for each possible allophone. The data will almost certainly be essential for any practical system. (Holmes, 1988:172f.)

This situation has not changed fundamentally. But current data-driven approaches to speech recognition can be criticised from the linguistic point of view for the following reasons:

- Although there has been a general move to incorporate more linguistic knowledge into data-driven approaches, it is still assumed that recognition must be in terms of straightforward non-overlapping units (words, syllables, demisyllables, phonemes, diphones etc.).

- The data-driven models do not take human linguistic competence and performance into account.

- Data-driven models are more concerned with system performance than dealing with native speaker performance (i.e. robustness issues).

- Data-driven models are primarily concatenative, while recent developments in linguistic theory have shown this assumption to be too restrictive.

These criticisms are discussed in detail in the next section. Before doing so, however, it is necessary to introduce the notion of phonological parsing.

The phonological parsing task involves the assignment of a phonological structure to a phonetic representation. The input can be in terms of strings of phonemes or feature bundles, or as shall be seen in Chapter 7, it can be nearer to the signal in terms of autonomous phonetic events. The output of phonological parsing can be in terms of syllable structures or phonological word structures. The knowledge base consists of rules describing particular phenomena and the processing component interprets these rules with respect to the input to provide the output structures.

In the majority of phonological monographs little is said about phonological parsing. An exception to this is Kaye (1989) who explicitly discusses issues of parsing in his book on a cognitive approach to phonology. His basic argument is that phonology provides the basis for processing. He is critical of derivational generative phonology whose formalism allows unrestricted rule types. The derivational approach is not an appropriate basis for the development of phonological parsers. He states

> The parsing problem is then to take a phonetic input and match it against the list of lexical representations stored in the brain. In order to do this, the effect of the phonological processes must be undone. That is, we must perform the converse of a phonological derivation. Starting with a phonetic form we must deduce its source. What underlying form could have led to the production of this form? In the early 1970s, such parsers were constructed. The results were quite startling. Even with full knowledge of all the phonological processes involved, the program churned out hundreds of possible sources for relatively simple input forms. This is hardly a plausible model for human linguistic behaviour. (Kaye, 1989:147)

Throughout the book, Kaye (1989) advocates a more restricted feature-based nonlinear approach to phonological description. Although he does not provide any details as to how phonological parsing can be undertaken on this basis, he briefly discusses (1989:158f) how phonological representations (in particular syllable structures) can provide important cues for spoken language recognition.

The most cited work (although, strangely, not by Kaye, 1989) in the area of phonological parsing is Church (1983, 1987b). Church (1983, 1987b) presents a phonological parser for speech recognition which uses allophonic cues for the determination of syllable boundaries. Church's claim is that phonological analysis should not rely on invariant phonological information as its sole source of knowledge. Allophonic variation is shown to be advantageous for phonological parsing since the realisation of particular allophones of a language is strongly dependent on their position within the syllable. Allophonic cues are not merely *noise* as claimed by researchers involved in acoustic-phonetic analysis (e.g. Klatt, 1979:284) but a source of cues to structure. Allophonic information is described using phrase-structure rules describing phonological well-formedness (or phonotactics). Church's allophonic parsing system has two stages from phonetic input to the point where phonemic output is matched with a syllable dictionary. A phonetic feature lattice incorporating generalisations about allophones is input to a bottom-up chart parser. This chart parser, which works on a similar basis to the well-known Cocke-Younger-Kasami algorithm (cf. Aho & Ullman, 1972), provides the phonetic input with a syllable structure. A canonicaliser, which undoes the postlexical rules, then discards the allophonic information and outputs a phonemic feature lattice preserving the syllable structure which comprises the input to the lexical matcher. The input to the phonological parser assumes earlier segmentation by an acoustic front-end.

Since Church (1987b), allophonic variation has been taken into account in two further models. The first is the hybrid system for spoken language recognition presented in Allerhand (1987). Allerhand combines pattern recognition with a knowledge-based approach to speech recognition. His knowledge component also contains production rules for phonotactics. The phonological parser described in Carson (1988, cf. also Chapter 6 for an extended version of this parser) uses as its linguistic knowledge base a declarative finite state network representation of the allophonotactics (allophonic syntax) and the phonotactics of a particular language (in particular for English and German) in order to parse a phonetic sequence into phonological words. The phonotactic nets for English and for German are feature-based and serve as phonological word templates consisting of *nonreduced* and *reduced* syllables. The transitions in a phonotactic net consist of transduction relations which rely on a restricted form of unification and elementary operations over sets in order to express the correspondences between the allophonic and the phonological domains. This model also assumes segmental input.

Gibbon & Braun (1988a) present a prosodic parser within an architecture for speech recognition which allows parallel, sequential, incremental (PSI) or parallel, hierarchical, incremental (PHI) processing. Prosodic parsing in this sense refers to the parallel processing of both segmental and suprasegmental phenomena which occur on autonomous tiers. This approach uses finite state transducers to simulate feature modules such as acoustic edge detectors and contour detection, and parallel and cascaded transducers to simulate clusters of features and autosegmental tiers (1988a:202). The approach differs from Kay's (1987) single multi-tape transducer for autosegmental relations (cf. Chapter 3) in using a configuration of several transducers; formally, the individual transducers have much in common with the structure of Koskenniemi's (1983b) two-level transducers (cf. Chapter 3), and to a lesser extent with Hidden Markov transducers.

The importance of more detailed phonological information for processing has been recognised in the psycholinguistic community also. In connection with psychological aspects of spoken word recognition, Frauenfelder & Lahiri (1989) recommend a closer look at phonological processes. They state:

> The psycholinguists' ignorance of basic phonological concepts is reflected by their view of input and lexical representations and by the way they have ignored problems posed by phonological processes. Until recently most models of lexical access have assumed extremely simple input consisting of linearly concatenated segments with no internal structure.... Although models making these simplistic assumptions might account for the recognition of isolated words spoken in citation form, it is evident that they cannot explain how listeners recognize words in fluent speech. (Frauenfelder & Lahiri, 1989:320f.)

Their examination of various possible psychological accounts of how listeners deal with phonetic variability leads them to discuss the solutions offered by the allophonic parser (Church, 1987b), and by other models (e.g. connectionist models) which include prosodic or syllable structure.

The phonological parsers discussed in this section have not found recognition in the speech community. The main reason for this has been that all these approaches make the general assumption that an explicit segmentation into non-overlapping units has to be carried out before any type of linguistic processing can take place. However, it is the segmentation or classification task which proves to be the most difficult in spoken language recognition. It is claimed in this book that what is required is a computational phonological model which can cater for overlapping phenomena and delay a strict segmentation to a later stage. Autosegmental phonology provides an ideal basis for such a model. Kornai (1991) suggests a linear encoding of autosegmental phonology which can provide the basis for what he terms *structured hidden Markov models*. A detailed discussion of Kornai's encoding is presented in Chapter 3 of this book. The encoding is designed primarily for two-tiered representations and a generalisation to multi-tiered representations gives rise to a vast amount of redundancy. Kornai states that the model has not been fully implemented and therefore has not been incorporated into a speech recognition system. However, this approach could be an interesting prospect for efficient future spoken language systems (cf. in particular the discussion of Kirchhoff (1995b) in Chapter 3).

1.2 CRITERIA FOR PHONOLOGICAL MODELLING IN SPOKEN LANGUAGE

Before designing a computational model of phonology, it is necessary to establish adequacy criteria which determine its appropriateness to linguistic description and to the spoken language recognition task. It is this aspect which speech components of current spoken language recognition systems neglect. While such systems have placed much emphasis on system performance, very little attention has been paid to linguistic competence since this is an area which is not considered relevant for engineering methods. The claim is made in this book (in particular Chapter 2 and Chapter 5) that evaluation criteria for computational linguistic models must be primarily concerned with linguistic competence and that modelling linguistic competence can offer a solution to the *new word* problem.

Chomsky (1964) proposed levels of success for grammatical description as measured with respect to the linguistic competence of a native speaker of a language; these measures still provide a useful point of reference for discussion. Levels of success for grammatical description can be regarded as a scale ranging from weak to strong linguistic adequacy. The weak level of linguistic adequacy which must be attained by a grammar is *observational adequacy* whereby the grammar describes the observed (or actual) data of the language. A stronger level of linguistic adequacy is *descriptive adequacy* whereby the grammar must distinguish between well-formed (actual and potential) and ill-formed structures of the language in such a way as to reflect native speaker competence. The strongest level of linguistic adequacy is *explanatory adequacy* which is attained if a theory can be established which provides a principled explanation for the inclusion of well-formed structures and the exclusion of ill-formed structures from the grammar. These adequacy criteria are referred to collectively as empirical adequacy criteria in the remainder of the book.

These criteria alone are not sufficient for evaluating a computational linguistic model and since further evaluation criteria for linguistics models have been proposed (e.g. Shieber, 1986; Langer & Gibbon (1992)). It is emphasised in Chapter 2 that a representation formalism for phonological description must be computationally interpretable. It must have a formal syntax and semantics and must be restrictive enough, to allow the description of those and only those phenomena which correspond to the linguistic competence of the native speaker. Furthermore the model must take the results of psychological experiments into account, that is to say, current knowledge about the way in which humans process speech. This latter criterion has played a role in the selection of relevant levels of description and in the way in which processing of event structures is undertaken.

In connection with the linguistic competence of a native speaker, Katz & Fodor (1964) discuss a phenomenon which they term the projection problem. They state:

> A synchronic description of a natural language seeks to determine what a fluent speaker knows about the structure of his language that enables him to use and understand its sentences. Since a fluent speaker is able to use and understand any sentence drawn from the *infinite* set of sentences of his language, and since, at any time, he has only encountered a *finite* set of sentences, it follows that the speaker's knowledge of his language takes the form of rules which project the finite set of sentences he has fortuitously encountered to the infinite set of sentences of the language. A description of the language which adequately represents the speaker's linguistic knowledge must, accordingly, state these rules. The problem of formulating these rules we shall refer to as the *projection problem*. (Katz & Fodor, 1964:482)

Katz & Fodor (1964) go on to state that any system of rules which aims to solve the projection problem must reflect in some way the "compositional character of the speaker's linguistic skill" (1964:482). Although Katz & Fodor are referring in particular to the semantic component in natural language description, their argument applies to all levels of linguistic description. Gibbon (1990b) summarises one of the major problems in the processing of speech in terms of a *projection problem* at the phonetics/phonology interface: sounds and words are realised with different degrees of coarticulation (overlap of properties) in different lexical, syntactic and phonostylistic contexts and thus a segmentation into phonemes alone is too rigid in order to capture all variants. Furthermore, the set of possible words in natural languages, analogous to the set of sentences, is infinite. In fact, even finite subsets of these sets may be so large that a simple list is no longer tractable. This has so far proved to be a major problem for the simple concatenative word models of current speech recognition whether phoneme, demisyllable or word based.

Linguistic competence is an issue which has to be addressed by a computational phonological model for speech recognition. However, native speaker *performance* is a further issue which has consequences for phonological parsing in a spoken language system. In continuous speech, human performance is often defective, that is to say, it may contain hesitations, repairs, or even syntactically incorrect or incomplete utterances. Continuous input to a spoken language recognition system may be underspecified or noisy and the phenomena found may not correspond to the well-formed structures of the language. It is clear that in order to develop an adequate computational model of phonology, consideration must be given to robustness issues and the properties of human speech performance.

The goal of this book is to present a linguistically and computationally adequate model of phonological processing in speech recognition which also goes some way to fulfilling cognitive criteria. *Time Map Phonology* promises a principled and effective solution to the projection problem, starting from recent well-motivated developments in phonology such as autosegmental phonology (Goldsmith, 1986, 1990), articulatory phonology (Browman & Goldstein, 1986, 1989), underspecification theory (Archangeli, 1988; Keating, 1988) and phonological event theory (Bird & Klein, 1990). This approach utilises a flexible notion of compositionality based on underspecified structures with 'autosegmental' tiers of parallel phonetic and phonological events which avoids a rigid mapping from phonetic parameters to simple sequences of segments. In this manner, all information in the acoustic signal can be made available to phonological processing and indeed also to higher levels of processing. By allowing a representation in terms of overlapping events at the phonetics/phonology interface, a representation which is nearer to phonetics than the phoneme sequences of more traditional phonologies can be obtained. This approach takes temporal interpretation of phonetics and phonology explicitly into account. By distinguishing between specified and unspecified temporal constraints on phonetic and phonological representations, robust parsing is permitted. Furthermore, the arguments in favour of a nonlinear approach to phonological processing are supported by recent discussions in Gibbon & Braun (1988a), Frauenfelder & Lahiri (1989), Kaye (1989) and Kornai (1991).

1.3 OVERVIEW OF THE BOOK

This chapter has seen a brief introduction to the problems facing an adequate computational model of phonological description which can find application in a spoken language recognition system. The following chapters address these issues systematically, developing a computational model which goes a long way to fulfilling the proposed adequacy criteria, which offers a solution to the projection problem in spoken language processing and which has been integrated into a spoken language recognition system. The book is structured in the following manner:

In Chapter 2 phonological description is introduced within the context of computational linguistics. As with computational syntactic and semantic descriptions, an important distinction is drawn between declarative and procedural aspects of phonological description, and declarative adequacy criteria for a computational model of phonology are proposed. It is shown that much of phonological description does not make this distinction between declarativity and procedurality explicitly. *Time Map Phonology*, a multi-linear computational phonological model based on events which serves as a basis for the rest of this book is then introduced. In this chapter, also, the domain of description is defined.

Chapter 3 is concerned with finite state approaches to linear and multilinear phonological description. This chapter aims to give an overview of the main developments in finite state phonology and discuss the role they can play in spoken language processing. In particular, just recently, there have been a number of developments in connection with finite state descriptions of autosegmental or multilinear phonology which are reviewed here. This chapter introduces the finite state device which is used for constraint-based event processing later in the book.

In Chapter 4, the motivation for the event concept in *Time Map Phonology* is discussed. Events are defined according to two domains: the temporal domain and the knowledge domain. Following Gibbon (1992d), three temporal domains which are relevant for phonological processing in a spoken language system are introduced. These are the absolute time domain T_{abs} (or utterance time), the relative time domain T_{rel} (or structural time) and the category time domain T_{cat}. A distinction is made between events in T_{abs} known as absolute-time events and events in T_{rel} known as relative-time events. Axioms governing the temporal relations between absolute-time events and relative-time events are defined. The knowledge domains relevant for phonological parsing in a spoken language recognition system are defined as the acoustic domain $K_{acoustic}$, the phonetic domain $K_{phonetic}$ and the phonological domain $K_{phonological}$.

Chapter 5 presents a discussion of linguistic adequacy criteria for phonotactic description and demonstrates to what extent phonotactic description formalisms fulfill these criteria. Two linguistically adequate phonotactic descriptions are presented which serve as knowledge components for the phonological parsers discussed in Chapters 6 and 7. Two network representation formalisms for phonotactic description are presented. The first of these representation formalisms allows for a feature-based segmental description of phonotactic combinations in a language. The second formalism allows for a new event-based, autosegmental description of the phonotactics of a language. Examples are provided for German syllables.

In Chapter 6, a segmental constraint-based approach to phonological parsing is illustrated which shows how underspecified segmental allophonic input can be enhanced and analysed using phonotactic descriptions and redundancy rules. This an extended version of the parser described briefly above (Carson, 1988). In addition, it is shown how speech variants can be described and processed in segmental terms. This model does not take temporality explicitly into account, however, but rather serves as basis for comparison with the nonsegmental event-based phonological parser described in Chapter 7.

Chapter 7 presents a constraint-based event processor is described which offers a solution to the projection problem in spoken language processing. This processor avails of many of the features of the segmental parser in Chapter 6 but assumes two temporal domain mappings ($T_{abs} \rightarrow T_{rel}$ and $T_{rel} \rightarrow T_{cat}$) within a particular knowledge domain. The operational semantics for each of these mappings are presented in terms of a virtual machine. In the final section of this chapter, questions of system architecture for a spoken language recognition system are discussed.

In Chapter 8, an interactive incremental architecture which accommodates linguistic modelling below the word level using a lattice event parser is introduced. Two example event-based phonological parsers are presented: COPPA+SILPA and an extension of this parser SILPA2. Both parsers use the operational semantics described in Chapter 7. Furthermore, SILPA2 has been provided with a parametrisable development environment for examining the extent to which linguistically significant issues such as linguistic competence and linguistic performance can play a role in the spoken language recognition task.

In Chapter 9 a new approach to the diagnostic evaluation of linguistic components of a spoken language recognition system is presented. This approach serves as the basis for the evaluation of the constraint-based event processor in a phonological parsing task for spontaneous spoken dialogue.

Chapter 10 concludes the book with a review of the goals, of what has been achieved and of the problems which still remain open.

2

PHONOLOGY AND COMPUTATION

Descriptive linguistic theories have been greatly influenced by computational linguistics. In computational linguistics, a fundamental distinction is made between the *declarative* and *procedural* aspect of a computational model of linguistic description. The primary concern of computational linguistics is with the declarative aspect, with formal, computationally interpretable representations of linguistic descriptions. The procedural aspect, the computational processing of the model in terms of an algorithm, is considered to be a separate issue. The influence of computational linguistics has been most obvious in the area of syntax and semantics where the search for a computationally interpretable grammar formalism has led to the introduction of unification-based grammars which in turn have led to new syntactic and semantic theories. This approach is primarily declarative in that structures are characterised by partial information and mutually independent constraints on well-formedness and is procedurally neutral in that no reference is made as to how the constraints should be applied. The trend in computational linguistics is therefore to design a declarative, processor-independent, monotonic grammar formalism which makes no commitment to a particular procedural interpretation with respect to analysis or generation. A declarative linguistic description allows for many different processing models and is not committed from the outstart to any particular one.

Phonology has been less influenced by computational linguistics in the past. New phonological theories have arisen as a result of wanting to describe new phenomena rather than as a result of the need for computational interpretability. The generative derivational approaches to phonology are inherently procedural in that they define algorithms which specify the order and/or the direction in which rules are applied to strings. A move in the direction of declarativity was taken implicitly by the proponents of Natural Generative Phonology (e.g. Vennemann, 1972; Hooper, 1976) where extrinsic rule ordering was discarded and generalisations were assumed to be natural and only applicable to a surface form. Rule ordering was intrinsic, that is to say, dependent on the data and on general principles. It is only in recent years, however, that phonologists have consciously focussed on the computational interpretability of phonological grammar formalisms. The explicit consideration of the declarative/procedural distinction has led to a new approach to phonological description which falls under the rubric of *declarative phonology* (cf. the collection of papers in Bird, 1991, in particular the article by Scobbie for an overview of declarative perspectives in phonology and more recently the paper by Scobbie et al., 1996).

That the declarative approach, which has now become standard in syntactic description, is relevant for phonology is somewhat controversial in the world of theoretical phonology, however. Bromberger & Halle (1989), claim in their article entitled "Why phonology is different", that although the declarative approach might be appropriate to syntax, there are certain phonological phenomena which can only be explained by rule order. They quote a phenomenon (the *writer/rider* example) which has now become a classical example in the declarative/procedural debate. They also make clear their commitment to the generative procedural interpretation. It will be necessary to return to their example and to their arguments in favour of this interpretation in the next section.

The next section of this chapter is concerned with the issues which form the basis of the declarative/procedural controversy in phonology. They are discussed in general terms and no commitment as to the meaning of declarative adequacy is made until Section 2.2 where adequacy criteria for phonological description languages are proposed. In Section 2.3, several existing descriptive phonological theories are then evaluated with respect to these adequacy criteria in the context of phonological modelling in the spoken language domain. In Section 2.4 the choice of the event-based phonological description language for *Time Map Phonology* is motivated.

2.1 PROCEDURALITY IN PHONOLOGICAL THEORIES

This chapter examines those aspects of phonological theory which have led to the claims that it is inherently procedural. No conclusions are reached in this section as to what constitutes a declarative phonology. Rather, this issue is postponed to Sections 2.2 and 2.3 where declarative adequacy criteria for the evaluation of phonological descriptions are discussed. Inherent procedurality has two extremes: processes of rule application may be part of a grammar, or grammar rules may be an integral part of a processor for the rules. The latter is, in general, an implementational issue and is of no immediate concern here. The former, however, is frequently encountered in phonological theories, mainly arising from the fact that these are committed primarily to a generative interpretation. There are three main issues in phonological theory which may be considered inherently procedural: rule ordering, rule application and the derivation of surface representations.

Generative phonological theory as described by Chomsky & Halle (1968) is derivational. It assumes that a linear surface phonetic representation is derived from an underlying fully-specified phonological or underspecified lexical linear representation. There are two types of rule in linear theory: phonological rules and redundancy rules. A phonological representation is derived from a lexical representation by applying redundancy rules. A phonetic representation is derived from a phonological representation by applying phonological rules. Phonological rules have the following general properties:

- They are nonmonotonic in that they can alter representations by deleting and inserting segments.
- They can alter the inherent structure of a segment by changing the values of a feature, in effect swapping segments.
- They are formulated as unrestricted rewrite rules.

Redundancy rules have a different status from phonological rules.

- They specify both segmental and sequential well-formedness conditions for phonological representations.
- They can be regarded as logical implication rules.
- They incorporate the notion of markedness (lexical default specifications) and are therefore not strictly monotonic.

Monotonicity is discussed again in Section 2.3 in connection with default rules for underspecification.

There are a number of properties of phonological theories which arise as a direct consequence of the derivational approach to phonology. One of these properties which may be considered to be inherently procedural in phonological theory is the order of application of phonological rules in derivations. An example which is often quoted in favour of rule ordering is the *writer/rider* example. A number of formulations of this neutralisation phenomenon of Canadian dialects are found in the literature but they all serve to illustrate the need for rule ordering. In these dialects, the distinction between /t/ and /d/ is neutralised and a vowel is lengthened if followed by a voiced consonant (or raised if followed by a voiceless consonant). The relevant phonological rules are:

$$V \rightarrow [+long]/\_\_\begin{bmatrix} C \\ +voice \end{bmatrix} \tag{2.1}$$

$$V \rightarrow [-low]\,/\_\_\begin{bmatrix} C \\ -voice \end{bmatrix} \tag{2.2}$$

$$\left\{ \begin{array}{c} t \\ d \end{array} \right\} \rightarrow \text{ɾ}/V\_\left[\begin{array}{c} V \\ -stress \end{array} \right] \quad (2.3)$$

In these rules, V and C stand for vowels and consonants respectively. Rules (2.1) and (2.2) are alternatives; which formulation is considered more appropriate is not at issue here. The first rule states that a vowel is lengthened before a voiced consonant and the second rule states that vowels are raised when followed by a voiceless consonant. The third rule applies together with either rule (2.1) or rule (2.2); it states that an apical plosive becomes a flap if preceded by a vowel and followed by an unstressed vowel.

Taking for example the rules (2.1) and (2.3), before an extrinsic ordering is imposed on these rules, two derivations are possible. (Similar derivations can be found in Schane (1973) for example).

Derivation 1:

	write	*ride*	*writer*	*rider*
Underlying form	#raɪt#	#raɪd#	#ra ɪt+ər#	#raɪd+ər#
Rule (2.1)	—	#raɪːd#	—	#raɪːd+ər#
Rule (2.3)	—	—	#raɪɾ+ər#	#raɪːɾ+ər#
Derived form	[raɪt]	[raɪːd]	[raɪɾər]	[raɪːɾər]

Derivation 2:

	write	*ride*	*writer*	*rider*
Underlying form	#raɪt#	#raɪd#	#raɪt+ər#	#raɪd+ər#
Rule (2.3)	—	—	#raɪɾ+ər#	#raɪɾ+ər#
Rule (2.1)	—	#raɪːd#	#raɪːɾ+ər#	#raɪːɾ+ər#
Derived form	[raɪt]	[raɪːd]	[raɪːɾər]	[raɪːɾər]

However, only derivation 1 provides the correct results. Derivation 2 does not distinguish between the realisations of *writer* and *rider*. The rules must be extrinsically ordered: rule (2.1) before rule (2.3). It is interesting to note that there are, in fact, dialects which exhibit the forms generated in derivation 2. The second formulation of this phenomenon is given by the rules (2.2) and (2.3). The following derivations are possible.

Derivation 3:

	write	*ride*	*writer*	*rider*
Underlying form	#raɪt#	#raɪd#	#raɪt+ər#	#raɪd+ər#
Rule (2.2)	#rɔɪt#	#raɪd#	#rɔɪt+ər#	#raɪd+ər#
Rule(2.3)	—	—	#rɔɪɾ+ər#	#raɪɾ+ər#
Derived form	[rɔɪt]	[raɪd]	[rɔɪɾər]	[raɪɾər]

Derivation 4:

	write	*ride*	*writer*	*rider*
Underlying form	#raɪt#	#raɪd#	#raɪt+ər#	#raɪd+ər#
Rule (2.3)	—	—	#raɪɾ+ər#	#raɪɾ+ər#
Rule(2.2)	#rɔɪt#	#raɪd#	#raɪɾ+ər#	#raɪɾ+ər#
Derived form	[rɔɪt]	[raɪd]	[raɪɾər]	[raɪɾər]

Linear generative phonology claims that the extrinsic ordering of these rules which states that rule (2.2) applies before rule (2.3) is the only way to produce the correct results. It will be seen below, however, that there are other ways to describe this phenomenon without resorting to ordered rules.

The *writer/rider* example uses a basic stipulated linear application of the rules (i.e. the rules are applied one after the other according to the ordering). This example also illustrates a number of rule ordering types. *Feeding order* is the order in which the output of one rule creates an environment for the next rule. This ordering is evident in derivation 2. The order of derivation 1 is known as bleeding order. *Bleeding order* is the order in which the output of one rule removes an environment for the next rule. Extrinsic (i.e stipulated) rule ordering is essentially procedural.

A further criticism of the approach of linear generative phonology is the intermediate levels of representation which are generated by individual rules. These intermediate representations may not correspond to any phonetic realisation (i.e. are not well-formed in this sense) and therefore they have no linguistic status. This fact alone is not procedural, but it is an undesirable side-effect of the commitment to a derivational approach.

There is another aspect of the generative approach to phonological theory which is procedural, namely rule application. Rule application refers to the way in which an *individual* rule is applied to a representation. The basic application principle of generative phonology is *linear application* as demonstrated in the *writer/rider* example. There are other rule application principles which have arisen in connection with both ordered and unordered rules. They are all inherently procedural principles because they each specify a procedure which is independent of the main principles of derivation (e.g. left-to-right, cyclic). In the generative approach, they are part of a theory and not part of a processor which interprets descriptions of the theory.

A selection of rule application principles are given below:

- *The transformational cycle*: Phonological rules apply in linear sequence to each constituent of the morphosyntactic structure beginning with the smallest and proceeding to successively larger constituents until the maximal domain of phonological processes has been reached. (cf. Chomsky & Halle, 1968: 349).

- *The iterative application principle*: a principle for extrinsically ordered rules. Given an input string (D) to a rule of the form $\alpha \rightarrow \beta \ / \ \gamma\_\_\delta$, identify all segments which meet the conditions specified in the rule and carry out the changes simultaneously generating an intermediate derivation D_i. With D_i as the input string, reapply the rule. This process is repeated until the rule can no longer be applied non-vacuously. The next rule in the ordering is then selected. This application principle permits feeding applications. (cf. for example Kenstowicz & Kisseberth,1979: 325).

- *The directional-iterative principle*: a principle for extrinsically ordered rules. A rule of the form $\alpha \rightarrow \beta \ / \ \gamma\_\_\delta$ can be applied to an input string (D) either iteratively left-to-right or iteratively right-to-left. Left-to-right application involves examining the leftmost segment of the input string, if the conditions specified in the rule are met, apply the rule generating the derivation D_{i+1} and take this as the input for the next application, otherwise take D_i. Move to the next segment of the input string and repeat the process until the rightmost segment has been examined. The next rule in the ordering is then selected. The right-to-left principle is applied analogously. (cf. Kenstowicz & Kisseberth, 1979: 326).

- *The simultaneous application principle*: a principle for unordered rules. Given an input string, the underlying form (U), to a rule of the form $\alpha \rightarrow \beta \ / \ \gamma\_\_\delta$, all αs in the context $\gamma\_\_\delta$ are converted to β regardless of whether they remain in that context after the rule has been applied elsewhere in the string. The original input string (U) determines the application of the rule to any point in the string. Bleeding or feeding is prohibited. Rules can only apply to the underlying form and therefore there are no intermediate levels of derivation. (cf. for example, Chomsky & Halle, 1968:344).

- *The random sequential rule application principle*: a principle for unordered rules. Given an input string to a rule of the form $\alpha \rightarrow \beta \ / \ \gamma\_\_\delta$, the rule is applied in any derivations in which its conditions are met. Note that this differs from the iterative application principle in that individual rules are not subject to ordering. This is also known as the free reapplication hypothesis. (cf. Kenstowicz & Kisseberth, 1979: 307)

- *The elsewhere condition*: a principle for unordered rules. This principle states that if two rules can apply to the same input, the more specific rule applies first blocking the application of the more general rule. (cf. Kiparsky, 1982). This is a very general default principle.

These rule application principles describe processes (e.g *bottom-up*, *breadth-first*, *left-right*) which should not be part of the description in a declarative approach. However, it is not only linear generative phonology which exhibits such features. Nonlinear phonologies, although their primary concern is representation, maintain this procedurality (e.g. directionality in construction of metrical feet or association of tone). They also belong to the generative tradition. The commitment to a generative interpretation has been reinforced more recently by Bromberger & Halle (1989) with the argument that phonological description represents the permanent memory of a speaker and must therefore come before the phonetic implementation as an utterance. Phonological terminology is itself frequently procedural and implies nonmonoticity; for example, *resyllabification*, *syllable contact repairs*, *reorganisation* can all only refer to *processes*, a procedural concept. The review of various rule application principles illustrates how the same conclusion is drawn in a number of different ways; the phonetic interpretation of a phonological form can be attained using different application principles. The fact that the rules of phonology are too procedural is frequently criticised in the literature (cf. for example, Bleiching, 1991; Broe, 1991; Carson-Berndsen, 1990, 1991; Coleman, 1990, 1992 who have shown how many of the procedural rules of phonological theories can be formulated declaratively).

From the point of view of computational phonology, a theory which allows a procedural interpretation only in terms of generation, mapping phonological to phonetic interpretations, is too restrictive, since a computational model of phonology should be relevant for both generation (in a more procedural sense) and analysis. It is possible to show that the inherently procedural properties of generative phonologies can be at least reduced, if not eliminated altogether. A theory must be restricted to the description of those rules of a phonological grammar which constrain a well-defined representation. Rule ordering should not be an issue, but follow from the ordering of well-motivated levels of representation. A direct consequence of this view is that there should be no derivations and therefore no controversial intermediate representations. Linguistically relevant levels may either be related by a direct mapping (as opposed to the composite function of the derivational approach) or there should only be one, *surface*, level of representation.

With respect to rule order, Karttunen (1991) demonstrates how the *writer/rider* example can be described without extrinsic ordering using unordered two-level rules of the form:

$$\text{aɪ} : \text{ɔɪ} \Leftarrow \_ - V : \tag{2.4}$$

$$\text{t:ɾ} \mid \text{d:ɾ} \Leftarrow V\_V \tag{2.5}$$

In this notation, the ':' distinguishes between levels. Rule (2.4) defines the correspondence between the *lexical* aɪ and the *surface* representation ɔɪ which is conditioned by the *lexical* environment (_-V: means 'followed by a voiceless segment at the lexical level'). Rule (2.5) describes the flapping of either ('|')

lexical /t/ or /d/ between vowels at both levels. In line with Kaplan & Kay (1981, 1994), it is possible to show that two individual finite-state transducers for these rules can be composed into a single finite-state transducer which maps between a lexical and a surface level of representation without recourse to intermediate levels of derivation. An example of this approach will be discussed in more detail in connection with finite-state techniques in Chapter 3.

Criticising the formulation of the *writer/rider* example in terms of rules (2.1) or (2.2) and (2.3) for being descriptively inadequate with respect to the actual data, Bird (1990) claims that this phenomenon should be described non-segmentally accounting for the temporal coordination of articulation. Bird's approach is to adopt a description/object distinction in which the representation/rule distinction has no place. The description/object distinction is a between linguistic types and token realisations, and closely related to the permanent memory/articulation and audition distinction proposed by Bromberger & Halle (1989:53) although they do not interpret it in the same way as Bird. For Bird, rules are also partial descriptions of phonological objects. They are interpreted as logical implication. Therefore, they do not alter descriptions by deletion, insertion or feature value changing, but rather they *constrain* the set of possible objects and most importantly they are well-formed. He proposes a declarative phonological model and provides a formalisation of autosegmental phonology and uses an event logic to incorporate the temporal dimension into the description.

In this section, inherently procedural properties of phonological theories have been sketched. Less space was reserved in this section for the declarative alternatives as these receive more detailed coverage in the chapters to come. An event-based formalisation of autosegmental phonology provides the basis for *Time Map Phonology*. It is assumed that, in general, there exist partial descriptions of various linguistic objects at the various linguistic levels. This notion is carried further in this book by providing partial phonological descriptions of phonological objects and partial phonetic descriptions describing phonetic objects. The next section is concerned with declarative adequacy criteria for phonological description languages (or theories). It is then demonstrated that the formalism of *Time Map Phonology* fulfils these criteria.

2.2 ADEQUACY CRITERIA FOR PHONOLOGICAL DESCRIPTION LANGUAGES

The fundamental assumption in computational phonology is the separation of the declarative and the procedural. A computational phonological description should be relevant for both analysis and generation and should not commit itself to either. This can only be done if all inherently procedural notions are removed from the description, if there is no specification of the way in which rules should be applied and if the main emphasis is on well-formedness of description. In this sense, phonological rule systems are considered here to be systems of constraints which are responsible for monotonic specification of well-formed descriptions. Nothing is said about the order in which these rule systems should be applied (cf. Pollard & Sag, 1987:7). It will be seen, in the next section, that most of theoretical phonology has committed its descriptions to the notion of generation, and since speech recognition is obviously committed to the notion of analysis, it has been unclear in the past where the common ground lies.

The diversity of description languages as used in phonological theories makes them difficult to evaluate as suitable formalisms for computational models. This is one of the reasons that modern phonological description has played practically no explicit role in spoken language systems. In order to remedy this situation, it is necessary to establish adequacy criteria for phonological modelling in the spoken language domain. Such adequacy criteria take into account empirical issues with respect to a particular linguistic domain and representational issues with respect to some formal or informal notation. On this basis a distinction is made between declarative adequacy criteria and procedural adequacy criteria (cf. also the adequacy criteria for lexical description given in Gibbon, 1992a; Langer & Gibbon, 1992 and Gibbon & Langer, 1992). Until recently, phonologists have been more concerned with declarative empirical adequacy criteria and speech engineers have been more concerned with procedural adequacy criteria. Computational phonology claims, however, that only a model which is both declaratively and procedurally adequate is capable of providing new insights to both phonology and to spoken language.

This section is primarily concerned with declarative adequacy criteria. Procedural adequacy is concerned with criteria for efficient computational representation and system performance. Computational phonology is concerned with these issues to the extent that a computational phonological model, which has a specified syntax, theory of inference and declarative semantics, should be provided with a procedural interpretation (operational semantics) in terms of an abstract machine specification. The implementation, in terms of a specific machine, is often a secondary issue from the point of view of computational linguistics and serves mainly as a test for the abstract model. Procedural adequacy criteria will be revisited in more detail in Chapter 7 in connection with an implementation of a phonological model for spoken language processing.

Declarative adequacy criteria are of theoretical interest as part of an empirical methodology for the development of descriptive computational linguistic theories. The classification of declarative adequacy criteria involves the notion of *empirical* adequacy (for the linguistic knowledge domain) and *representational* adequacy for the formalism used to describe the knowledge in the domain. Empirical adequacy is already familiar in linguistic description whereas representational adequacy has arisen in connection with computational theories of language. In the discussion which follows, declarative adequacy criteria are not defined with respect to any particular empirical phonological domain; rather this is given as a parameter P. The reason for this will become clear in Section 2.3. Let it suffice to say that the term *phonological* has been used to apply to differing domains of generalisation sometimes termed lexical, postlexical, phonological, morphophonological etc. This makes the provision of suitable evaluation criteria difficult since different descriptions refer to different domains. The notion of a domain in this sense is directly connected to the notion of level as described in the previous section.

The following levels of empirical adequacy (based on the classification given in Gibbon & Langer (1992) modified from Chomsky (1964)) should be taken into consideration for the evaluation of description languages in an empirical phonological domain P:

- *Observational phonological adequacy*: the ability to represent *actual*, observed phonological expressions in a given empirical phonological domain P.
- *Descriptive phonological adequacy*: the ability to generalise beyond *actual* phonological expressions in the domain P, making predictions, on the basis of the observed expressions, about *potential* compositionally well-formed phonological expressions in P, thereby distinguishing accidental from systematic gaps of P.
- *Explanatory phonological adequacy*: the ability to provide a principled reason for selecting the optimal description with respect to P.

Descriptive adequacy is taken as the minimum requirement for further evaluation of phonological description languages, although it will be seen in the next section that this is exactly the criterion which current models of spoken language lack. A descriptively adequate phonological description language must look at which type of generalisations are allowed and required in order to predict *potential* expressions which are valid in the domain. This point is taken further in connection with particular phonological description languages in the next section.

In addition to criteria for empirical adequacy, other evaluation criteria for computational linguistic formalisms have been proposed. These criteria concern mathematical adequacy and notational adequacy (cf. Gazdar, 1985) or computational effectiveness, expressiveness and linguistic felicity (cf. Shieber, 1986). Mathematical adequacy is considered to be a notational requirement and not a processing requirement. The notion of restricted expressivity is also not directly related to procedural adequacy. It refers to the grammar formalism with respect to the phenomena to be described. There may be reasons for choosing a more powerful mechanism at the implementational level (cf. Berwick & Weinberg, 1982).

A formal approach to phonological description is taken in this book which takes many of its ideas from model theory (cf. also Bird, 1990). In order to develop a computational model for a theory, a model-theoretic interpretation (declarative semantics) is required. In addition, a theory should be considered from a proof-theoretic point of view in terms of a procedural semantics which defines the ways in which new well-formed expressions can be inferred from the basic axioms. These criteria are referred to collectively here as representational adequacy. The representational adequacy criterion for notation used for a phonological description language refers to the following complex notion:

- *Representational adequacy*: the ability to express all and only well-formed objects of a domain P in a notation appropriate to the domain P which has a clearly defined formal syntax and semantics.

As with empirical adequacy, it would also be possible to define types of representational adequacy. It may be the case, for example, that the representation formalism restricts the class of objects which can be expressed without using a notation which is appropriate to the phonological domain P. On the other hand, the notation may be phonologically appropriate (although this may be a matter of opinion, cf. Shieber, 1986) but not have a clearly defined formal syntax and semantics. It is also possible that the notation has a clearly defined formal syntax and semantics but is unrestricted from the point of view of expressive power, that is to say, is not mathematically adequate (in the sense of Gazdar, 1985). A division into relevant levels of representational adequacy is problematic since it is not clear which level is more adequate than another (this criterion is not really scaleable in this sense).

In the next section, representational adequacy is taken as the major criterion for a phonological formalism and in the discussion, it is indicated to what extent a particular phonological description language fails to fulfill the representational adequacy criterion.

The declarative adequacy criteria defined in this section are closely related to the criteria of epistemological adequacy, logical adequacy, algorithmic adequacy, ergonomic adequacy and psychological adequacy discussed in Sagerer (1990) in connection with knowledge representation formalisms for speech understanding systems. In connection with the linear encoding of autosegmental phonology, Kornai (1991:23f) presents criteria for the 'optimal code'. These criteria concern computability, invertibility, iconicity and compositionality. While these criteria are indirectly related to the notion of representational adequacy presented here, they do not take the notion of a domain explicitly into account. It would be difficult to claim that Kornai's notation is *appropriate* to a linguistic domain (cf. the more detailed discussion in Chapter 3). However, it is not intended as an external notation for linguistic description, but as an internal representation for efficient processing of autosegmental representations.

2.3 HOW ADEQUATE ARE PHONOLOGICAL THEORIES?

Now that declarative adequacy criteria have been defined, it is the task of this section to relate phonological grammar formalisms to these criteria in order to justify the choice of a particular formalism for computational phonological description. It was seen in section 2.1 that generative phonology was considered to have many procedural properties which indicate already that it does not fulfill the criteria. However, it is necessary to pinpoint where the deficits lie. Descriptive adequacy is assumed to be the basic empirical level which formalisms suitable for computational phonological modelling must attain.

2.3.1 Linear Approaches

It was seen in Section 2.1 that although linear generative phonology made a distinction between representations and rules, the inherently procedural properties with respect to the application of the rules exclude this theory from fulfilling the declarative adequacy criteria required for computational phonological modelling. On the empirical side, the aim of linear generative phonology is obviously to develop a linguistically adequate theory and, for this reason, it must be at least descriptively adequate. However, if later developments concerning the description of suprasegmental phenomena are taken into consideration, it would seem that their descriptive adequacy must be questioned. Furthermore, proponents of Natural Phonology (cf. Stampe & Donegan, 1979) criticise the fact that linear generative phonology provides no way of falsifying the descriptive adequacy criterion as the theory is unrestricted. Kaye (1979: 111) in his discussion of linear generative phonology, goes as far as to state that "it was possible to write absolute nonsense using this theory". This is not to claim, however, that linear generative phonology is only observationally adequate.

The formal notation of linear generative phonology is given in Chomsky & Halle (1968:390ff). As far as notational adequacy in the sense of linguistic felicity is concerned, this matter is open to discussion. At the time linear generative phonology was developed, this, indeed, was a notation which seemed to allow the required generalisations. However, it is surprising, that in spite of the fact that Chomsky developed the formal foundations for defining the expressive power of grammar formalisms, mention is seldom made as to whether phonological rules are in any way restricted. Chomsky's syntactic studies had apparently led him to believe that Type 0 grammars were required at every level of description. However, unrestricted rewrite rules, which have Turing machine power, are an over-kill if phonological expressions are examined more carefully. Johnson (1972) examined this issue more closely, noting that in contrast to syntax, phonological expressions can be analysed as regular expressions (i.e. finite state) and that various types of rule application could be mimicked by finite state transducers. It is this fact which allows Karttunen (1991) to respond to Bromberger & Halle (1989) by claiming that phonology is indeed different, although not in the sense that the latter authors intended, namely that phonology requires a procedural approach, but rather that the rules of generative phonology can be stated declaratively. The notion of regularity and finite state phonology are considered further in Chapter 3.

Lexical Phonology (Kiparsky, 1982) is closely related to linear generative phonology except that the lexicon is considered to be the major component rather than being an appendix to the grammar as was the case in linear generative phonology. From the point of view of empirical adequacy, it would seem to be on a par with linear generative phonology. It does assume separate representation levels or strata for morphological and phonological description. However, it is a segmental theory which still maintains the notion of derivations, rule ordering and most importantly cyclicity of rule application. It is inherently procedural and unrestricted and for this reason cannot be considered representationally adequate for computational phonological modelling, although a computational model of lexical phonology has been developed by Williams (1991). Some moves have been made to relate aspects of the theory to computational linguistics by demonstrating how non-monotonic feature changing rules can be eliminated using unification and generalisation (Wiese, 1990), but the latter are not considered to be part of the theory: even if they were, unification and generalisation alone do not make a theory computationally interpretable. The notation of lexical phonology does not have a clearly defined formal syntax and semantics.

2.3.2 Nonlinear Approaches

The next two theories to be discussed are nonlinear. Both Metrical Phonology (Liberman & Prince, 1977) and Autosegmental Phonology (Goldsmith, 1976, 1990), and also other hierarchical syllable-based approaches such as Kahn (1976), Clements and Keyser (1983) and Selkirk (1982), use graphical representations since their primary interest is in phonological organisation. The notations do not fulfill the representational adequacy criterion and are not used consistently by all authors (cf. the criticism of autosegmental notation in Bird & Ladd, 1991). Many of the conventions and rules employed in these theories can be reformulated in a declarative way but to the extent that these theories both have rules which 'reorganise' phonological descriptions, they still have inherently procedural properties. Autosegmental phonology has provided the basis, however, for recent developments in declarative phonology and it is discussed in this context with respect to representational adequacy below. From the empirical point of view, the claim to descriptive adequacy has been based on the description of 'hard cases' and not on the complete phonological description of a language. Autosegmental phonology was developed originally to cater for tonal phenomena which was a problem for traditional linear theories. Autosegmental phonology rejects the *absolute slicing hypothesis* (Goldsmith, 1976) - that speech can be exhaustively sliced into discrete linear segments of unordered feature bundles - and treats phonological representations as an *orchestral score*. An orchestral score is a multi-linear representation of separate tiers (e.g. tonal tier, nasality tier) each of which has its own synchronisation function (similar to the musical score for an orchestra). The segments on the individual tiers are called autosegments since they are independent of segments on other tiers. Autosegments are related to autosegments on other tiers by *association lines* which express coarticulations. Metrical phonology was originally developed as a theory of stress, and in this sense can be understood as taking over where autosegmental phonology leaves off. Both theories have been used to describe similar phenomena, however.

Another theory concerned with phonological organisation is Articulatory Phonology (Browman & Goldstein, 1986, 1989, 1990). This approach describes an utterance as an organised pattern of overlapping independent articulatory gestures. Browman & Goldstein provide a formalisation of their theory of organisation in terms of task dynamics and show how gestures can be modelled computationally within this framework for speech synthesis. Articulatory phonology rejects the notion of phonological processes such as deletion and insertion; rather these *processes* are explained in terms of gestural magnitude and intergestural overlap. This is the notion which has been adopted in other declarative phonologies. Ladefoged (1990) has criticised this model for two reasons. Firstly, since tiers are defined according to articulatory independence, the articulatory model would propose that speakers who show a dependency between particular articulations have a different phonology from those who do not show this dependency. He emphasises, however, that this is not a problem for the model; it just requires that more tiers be specified. His argumentation leads to the conclusion that articulatory independence should not be the criterion for defining tiers. The second criticism is, for Ladefoged, more serious in that it concerns phonological description as a discipline; the articulatory model has no way of describing phenomena such as vowel raising since it does not postulate an underlying lexical representation. The problem here is that it is not clear to which domain (or level) the model refers. This point was mentioned in connection with the empirical adequacy criteria in the previous section and is an issue which emerges in connection with all phonological theories discussed here (except Lexical Phonology to the extent that it explicitly separates its levels). It is possible to go as far as saying that these models are not descriptively adequate for a phonological domain P because it is not clear what P is. This issue is taken up again in Section 2.4.

2.3.3 Constraint-based Approaches

Two constraint-based approaches to phonological description which are committed to declarativity are Constraint-based Phonology (Bird, 1990, 1995) and Unification Phonology (Coleman, 1990, 1992). Both of these theories emphasise the need for computational interpretability. The basic claim in both of these theories is that rules are well-formed partial descriptions of phonological objects and there is therefore no distinction between rules and representations. Bird provides a formalisation of the notions of autosegmental phonology in terms of an event logic where association is interpreted, following Sagey (1988), as temporal overlap and shows how a phonological description can be constructed using constraint satisfaction. The interpretation of association differs from that of Sagey in that Bird assumes temporal overlap to apply to intervals rather than points. Coleman uses a somewhat different notion which is more in line with the Firthian prosodic phonology tradition. Both these theories postulate the temporal relations of overlap and precedence in a similar vein to articulatory phonology. It is fair to say that both of these theories are representationally adequate although paradoxically it is not clear to what extent constraint-based phonology is explicitly restricted. Recent work by Bird (Bird & Ellison, 1992) indicates that it can be formulated in finite-state terms. Coleman (1992) does not claim that his approach is finite-state but states explicitly that grammars in his unification phonology are restricted to context-sensitive power (without deletions) in the worst case. Both theories provide a criterion for testing (or falsifying). However, the domain problem seems to arise again here. It is not clear to which phonological domain P constraint-based phonology is addressed. There is no notion of a lexicon yet many of the examples shown describe morphophonological phenomena. Coleman does not refer to the lexicon explicitly and it would seem that his descriptions are extra-lexical. On the procedural side, Bird provides an operational semantics for his model and states that a pilot version of the system has been implemented (1990:73), although this is not described further. Coleman has implemented a phonetic interpretation model for speech synthesis (YorkTalk, cf. Coleman, 1992). The system uses a phonotactic parser to construct headed phonological structures which are then interpreted to get parametric phonetic descriptions which serve as input for a Klatt formant synthesiser (1992:25). A further study of formal aspects of autosegmental phonology is found in Attribute value phonology (Scobbie, 1991b). Scobbie emphasises the need for declarativity in phonological description and also proposes constraints on well-formedness. He does not discuss computational aspects of his theory explicitly, however.

Another theory which has emerged since the model proposed in this book was developed and which has since been very influential in the area of phonology, is Optimality Theory (Prince & Smolensky, 1993). This theory explores the notion that Universal Grammar consists of a set of constraints on well-formedness of representations. Individual grammars for specific languages define language-specific rankings and conditions governing violation of these constraints. The constraints of Universal Grammar are therefore not required to be mutually consistent. The basic idea of Optimality Theory is that each input representation is associated with a candidate set of possible analyses as defined by Universal Grammar. Constraint satisfaction based on the ranking of the universal well-formedness constraints and the conditions for conflict resolution allows the selection of the *optimal analysis*. "The heart of the proposal is a means for precisely determining which analysis of an input *best satisfies* (or least violates) a set of conflicting well-formedness conditions" (Prince & Smolensky, 1993:2). The grammar defines a pairing between underlying and surface forms. A function *Gen* generates possible candidate analyses and the function *H-eval* evaluates the output determining the degree to which the candidates satisfies the constraints; this they term the *Harmony* of the analysis. The application of Optimality Theory has been demonstrated on the phonological structure of several languages (e.g. Tashlhiyt Berber, Lardil) and indeed this approach is attractive. However it is unclear whether a restriction is imposed on the types of grammar which can be written or whether this is left to the phonologist. A clear distinction between the violation of what could otherwise be termed language-specific competence constraints (i.e. constraints on a native speakers knowledge of a particular language) and language-specific performance constraints (i.e. they way in which a native speaker actually performs during speech) would seem to be impossible with this approach. It is possible, however, to visualise the application of a similar notion of ranking to language-specific well-formedness constraints which could be used for defining the application of constraint relaxation in speech recognition. This notion will be returned to in Section 8.4.

2.3.4 Underspecification-based Approaches

A topic which has long played a role in connection with the evaluation of phonological theories is underspecification. Lexical representations are assumed to be nonredundant, specifying only information which is not predictable. The more underspecified the lexical description, the more highly valued the theory was considered to be. Predictable specifications were added to the descriptions by redundancy rules (cf. also Section 2.1 above). The extent to which phonological or lexical descriptions should be underspecified has changed over the years. Underspecification Theory, as it is now understood, is discussed in detail by Archangeli (1988). She distinguishes between segment-based inherent underspecification and contrastive underspecification, and feature-based radical underspecification. This latter type of underspecification goes further than contrastive specification by removing all redundancies allowing for a variety of representations providing a basis for the notion of preferability or *markedness* in Universal Grammar. An unmarked or default form is not specified. Underspecification phonology also differentiates between language-specific and universal defaults. The difference between theses two types can be formalised using the Generalised Phrase Structure Grammar notions of *feature cooccurrence restrictions* and *feature specification defaults* (as also done by Calder & Bird, 1991). Contrastive specification uses only feature cooccurrence restrictions and radical underspecification uses both of these devices. In declarative theories, underspecification is understood as partial specification or description. There has been much discussion as to whether the defaults of underspecification theory are suitable for a declarative phonological framework which is supposed to be monotonic (cf. Scobbie, 1991a; Coleman, 1992). Two views seem to be prevalent. On the one hand, this aspect of phonological description can be formulated within a nonmonotonic logic (cf. Calder & Bird, 1991) and therefore may be permitted in the declarative approach. On the other hand, Coleman (1992:8) states that the fact that defaults are often formalised using nonmonotonic mechanisms does not mean that they are inherently nonmonotonic; if defaults and nondefaults are mutually exclusive, they can operate monotonically. Other default approaches to phonological and morphophonological description using DATR are presented in Gibbon (1990a), Reinhard & Gibbon (1991) and Cahill (1990, 1993)). Gibbon presents a DATR implementation of nonconcatenative morphology and Reinhard & Gibbon for umlaut phenomena in German using what they term quasi linear precedence. Cahill (1990) presents a syllable-based morphological description of German. In the 1993 paper, Cahill introduces the notion of temporal specifications into the lexicon. Although this book advocates the notion of temporal specifications of phonological entities and how they can be processed in the context of spoken language recognition, it is questionable whether temporal specifications of the kind Cahill (1993) proposes are relevant for the lexicon or whether a relative-time specification would not be more suitable (cf. Chapter 4).

2.3.5 Engineering Approaches

Although this book is committed to a declarative constraint-based approach to phonological description, its overall theme is to show how a computational phonological model can be used for *processing* spoken language. It would, therefore, be a serious omission to ignore successful work in the area of speech recognition. However, since this book is more concerned with a general approach to phonological inference rather than procedurally optimised specific devices, traditional speech engineering methods are discussed briefly in terms of the extent to which they can be evaluated with respect to the declarative adequacy criteria for a computational phonological model presented in the previous section. The question of procedural adequacy of such approaches speaks for itself and is not an issue here. The rest of this section provides some indications as to why the further progress of such approaches will be limited in continuous speech recognition.

Considering, for example, the *hidden Markov model* technique, it is clear that this approach as it is employed in current speech recognition systems is only observationally adequate to the extent that models are trained with respect to some set of *actual* forms in a corpus. Test data differs an training data are distinct and both consist of actual forms. Recent developments in connection with HMMs for speech recognition have discarded the word as a primitive for modelling and use a more structured approach in terms of primitives such as diphones, triphones or demisyllables (cf. Schukat-Talamazzini et al., 1991). However, the generalisations made by such models are based on concatenation only and there is no way, without consulting a lexicon, of distinguishing well-formed expressions from ill-formed expressions. On this basis, *potential* forms cannot be described. Furthermore, there is no specific treatment of underspecified data. It is not possible to say very much about the representational adequacy of this approach; from a linguistic point of view there is no notation. However, as mentioned in Chapter 1, many parallels can be drawn between HMMs and finite-state models of phonology. These similarities are made explicit by the comparison given in Gibbon (1992c, Slides: Lexical Access 11ff). Recent developments in the area of phonological description can provide useful *a priori* information for the construction of Markov models. Kornai (1991) proposes an encoding of autosegmental phonology designed to serve as a basis for structured Markov models for speech recognition (cf. Chapter 3).

Turning now to examples of models which more obviously take structure in a phonological sense into account by using explicit network representations, it can be seen that such descriptions are more easily related to the declarative adequacy criteria. The use of networks in phonological description is appealing since they can be formalised in terms of very simple notations. A network representation formalism for phonological description is proposed and discussed with respect to the adequacy criteria in Chapter 5 of this book. Two examples of network models which have found application in the area of speech recognition are described in the remainder of this section.

The first model is the demisyllable classification approach (Schotola, 1984; Seck & Ruske, 1986) to word recognition. Syllables are classified according to the initial consonant clusters (icc) plus vowel and vowel plus final consonant clusters (fcc) as demisyllables (cf. Fujimura, 1976 where this notion was introduced into phonological description). For German, a table of 47 initial consonant clusters and 159 final consonant clusters is needed. This is missing some generalisation; furthermore, there is no psycholinguistic evidence for the demisyllable primitives (cf. Chapter 4). The demisyllable classification approach is based on a verification network which expresses the temporal ordering of the internal structure (the tactics) of the demisyllables. The verification network given in Seck & Ruske (1986) is obviously more than observationally adequate in that it can describe *potential* forms which are not in the observed data. However, due to the fact that the verification network does not explicitly express restrictions on permissible combinations, it overgeneralises and is therefore not descriptively adequate since it does not distinguish between accidental and systematic gaps. In fact a study of the phonotactics of German described in Carson-Berndsen et al. (1989) shows that although well-formedness of a syllable onset can be described using 60 initial clusters, this amount can be reduced to only 29 clusters if underspecified feature classes rather than phonemes are used.

The second example of the use of network representation for phonological description is found in the neural nets of the connectionist approach. An overview of the use of neural nets in spoken language (both analysis and generation) is presented in Kemke (1990; also Lippmann, 1989 in connection with speech recognition). In the analysis direction, these models have been used for phoneme and phoneme class recognition, for syllable and word recognition either holistically or as a multi-layered system (e.g the TRACE model, McClelland & Elman,1986). The problem with holistic neural net systems is the same as that noted for HMMs, namely that only *actual* forms can be simulated. Depending on how the neural net has been constructed, it may be the case that the more complex models are also only observationally adequate (although they do aim to provide a cognitive explanation for the phenomena they describe). This would be the case with a trained neural net which has no additional phonological structure since there will never be enough observed data to make simulations equivalent to formal observations in a linguistic theory (cf. Bird & Ladd, 1991:206). In order to be descriptively adequate, linguistically significant generalisations must be made over the *actual* forms (cf. the multi-layered model described in Schade, 1987 and Eikmeyer et al., 1991). Gibbon & Braun (1988b), in their discussion of phonological networks, illustrate how the local connectionist net describing syllable structure given in Schade (1987) can be decomposed into subnets which express significant phonological generalisations such as structural implications (expressed by control nodes) constituent structure and paradigmatic restrictions. Here again, phonological generalisations can provide useful *a priori* information for the construction of neural net models. A connectionist model of cognitive phonology designed within the generative paradigm, based on the ideas of Lakoff (1987), is the basis for the *many-maps architecture* (Wheeler & Touretzky, 1989). However, a neural net encoding of constraints on phonological well-formedness has remained a problem for this model (Wheeler & Touretzky, 1989:1; 1990). The model is discussed further in connection with finite state phonology in Chapter 3.

In spite of the fact that it is claimed above that traditional speech recognition methods are not adequate as computational phonological models, there is no doubt that these methods are very efficient at the performance level for very specific corpora. However, due to the fact that the generalisations made by these models are based on concatenation, they assume a rigid segmentation into non-overlapping units. In view of the long term aims of continuous spoken language recognition, a descriptively and representationally adequate phonological model which caters for compositionality and variability thus offering a solution to the projection problem at the phonetics/phonology interface is essential.

2.4 FEATURE REPRESENTATIONS IN A COMPUTATIONAL MODEL OF PHONOLOGY

This chapter has been concerned with the adequacy of phonological theories for computational modelling. The final section of this chapter motivates the approach taken in *Time Map Phonology* and relates it to the adequacy criteria proposed in Section 2.2. More detailed descriptions of particular aspects of the computational model of phonology are the content of the chapters to come.

Very little has been said so far about feature descriptions and although no preference for any particular feature set is expressed in this book, it is necessary to look at the objects which have to be described in order to motivate the chosen representation. In order to do this, some simple examples of plosive-vowel-nasal sequences have been chosen to illustrate the various representations. These representations are based on those given in Carson-Berndsen (1991). The segmental description given in Figure 2.1 is the starting point of the discussion.

k	a:	m
manner(plosive)	manner(vowel)	manner(nasal)
place(velar)	place(back)	place(labial)
phonation(voiceless)	phonation(voiced)	phonation(voiced)
	f1-position(very-high)	
	length(long)	
	openness(open)	
	roundness(nonround)	

Figure 2.1: Segmental description of the sequence [ka:m]

A segmental multi-valued feature classification (cf. Carson-Berndsen et al., 1989) is employed here in order to avoid criticism on the grounds that binary-valued distinctive features are not suitable for spoken language.

The main problem with this representation is that the features all use the same synchronisation function. The absolute slicing hypothesis (cf. 2.3) is assumed. However, it is not an unusual occurrence that a feature be present throughout the whole description. The feature *voiced* extends across two segments in this simple example. This synchronisation function is an artefact of the segmental approach and it is clear that acoustic signals cannot be split up into segments of this type.

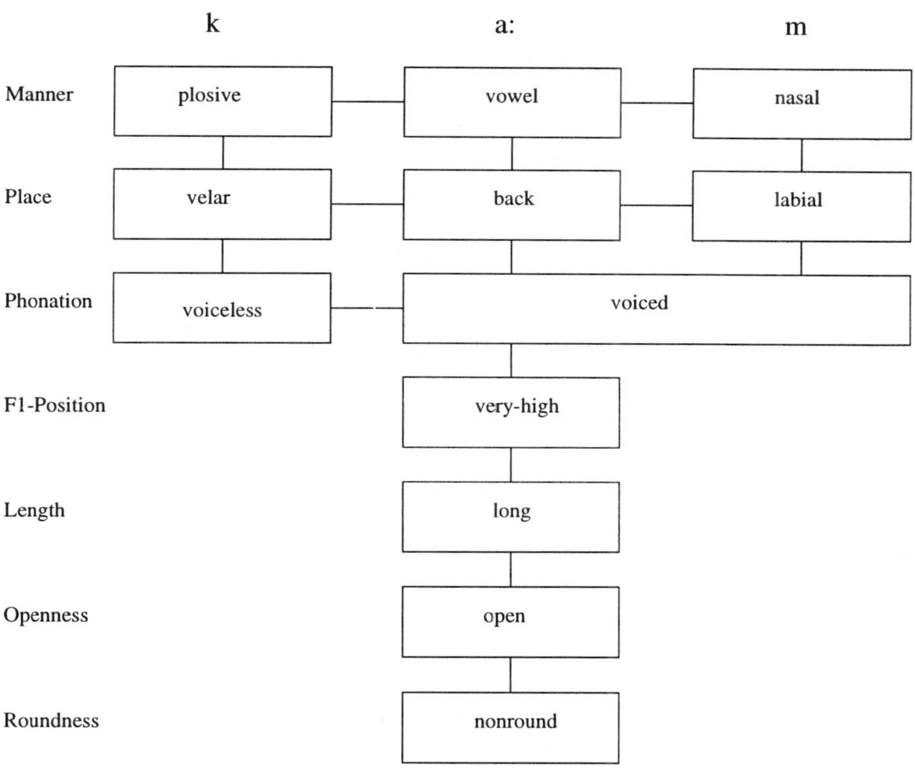

Figure 2.2: An 'autosegmental' representation of the sequence [ka:m]

On the other hand, if the features are regarded as independent entities, a representation which resembles the orchestral score of autosegmental phonology may be obtained (cf. Figure 2.2). The features indicate tier names and their values are the autosegments on the tiers. The horizontal lines indicate sequential order and vertical lines indicate coarticulation of features (association). This is a rather atypical autosegmental representation. Autosegmental representations typically have separate tiers for timing (or CV-skeleton), for tone or for nasality. In this description, however, there is a separate tier for every type of feature. Note that association may be between all tiers and not just between any two. This description can nevertheless be interpreted in the same way as an autosegmental representation. The motivation for this type of description, which is unusual for languages like English or German, is discussed in more detail below. A temporal dimension can be introduced if association is interpreted as temporal overlap (cf. Sagey, 1988; Bird & Klein, 1990) and the sequencing as temporal precedence. The durational aspect is lacking, however, and it is not clear from the diagram, that the synchronisation function can differ for each tier.

An articulatory gestural score, on the other hand does indicate to what extent individual gestures overlap. The gestural representation for the sequence [tɑ:n] is given in Figure 2.3 (based on the representations given in Browman & Goldstein, 1989). The articulators in the example are *velum (VEL), tongue body (TB), tongue tip (TT) and glottis (GLO)*. It is clear from this representation that the articulators can operate independently and yet be simultaneously active.

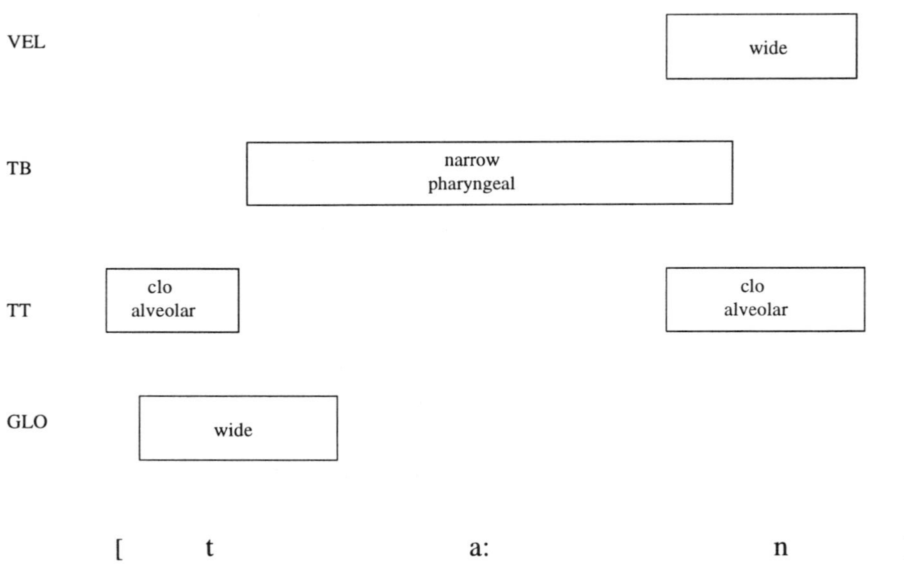

Figure 2.3: Articulatory gestural score for the sequence [tɑ:n]

In the work of Bird & Klein (1990), the relevant coarticulation and sequencing relations for autosegmental description are axiomatised in terms of an event logic. The details are not essential to the discussion in this section (but cf. Chapter 4); it is sufficient to say here that an autosegmental representation can be interpreted as an abstract description of temporal relations between autosegments. It is then possible, if required, to define the notion of a phonological segment, as a set of temporal relations between events. The example of the sequence [tɑ:n] in Figure 2.3 can also be interpreted as an autosegmental representation whereby a /t/ has been uttered if there is an overlap of alveolar closure and a wide glottis, an /ɑː/ has been uttered if there is a narrow pharyngeal gesture and a /n/ has been uttered if a wide gesture of the velum overlaps with an alveolar closure. Bird & Klein (1990:50) present this more formally. The following corresponds to their segment definitions:

$$t =< [i_1 : (closure, alveolar)]_{\mathrm{TT}}; [i_2 : wide]_{\mathrm{GLO}}; i_1 \circ i_2 >$$
$$\text{ɑː} =< [i_3 : (narrow, pharyngeal)]_{\mathrm{TB}} >$$
$$n =< [i_4 : wide]_{\mathrm{VEL}}; [i_5 : (closure, alveolar)]_{\mathrm{TT}}; i_4 \circ i_5 >$$

In this notation i_1.....i_n refer to intervals, the subscripts refer to the articulators and the $\circ$ symbol represents temporal overlap. The first line is equivalent to stating that a /t/ is articulated if there is an interval i_1 during which there are *closure* and *alveolar* gestures of the tongue tip and an interval i_2 during which there is a *wide* gesture of the glottis, and the intervals i_1 and i_2 overlap. Analogously, the phonological segment /t/, for example, in Figure 2.2 can be defined:

$$t = < [i_1 : plosive]_{\mathrm{manner}}; [i_2 : apical]_{\mathrm{place}}; [i_3 : voiceless]_{\mathrm{phonation}}; \\ i_1 \circ i_2, i_2 \circ i_3, i_1 \circ i_3 >$$

This interpretation allows for the fact that the intervals i_1, i_2 and i_3 do not necessarily begin and end exactly simultaneously. It is unnecessary to define segments under this interpretation. A complete phonological description can be provided without direct reference to phonological segments. A problem with this interpretation, as it stands, is that it is not clear how these abstract intervals relate to absolute time and duration. This book is concerned not only with abstract temporal relations but also with phonetic reality. In other words, it will be shown how phonological descriptions (*types*) can be related to a *token* utterance. In particular, it is concerned with phonological descriptions of phonological objects (or events) and phonetic description of phonetic objects (events). Both levels can be described declaratively using an event logic.

In order to illustrate a phonetic token description, the sequence [pɑːm] is presented, this time using the phonetic *segment type features* of Fant (1973:27ff). This example is based on one of the examples chosen by Browman and Goldstein (1989). A recording of a token utterance of /pɑːm/ was made and a signal annotation was provided (using the SAMLAB software, Braun, 1991). The annotation is given in Figure 2.4. According to this annotation, the token utterance /pɑːm/ is 519ms in length. The segments have the following lengths: /p/ : 94.25ms; /ɑː/ : 298.75ms and /m/ : 126ms. The temporal

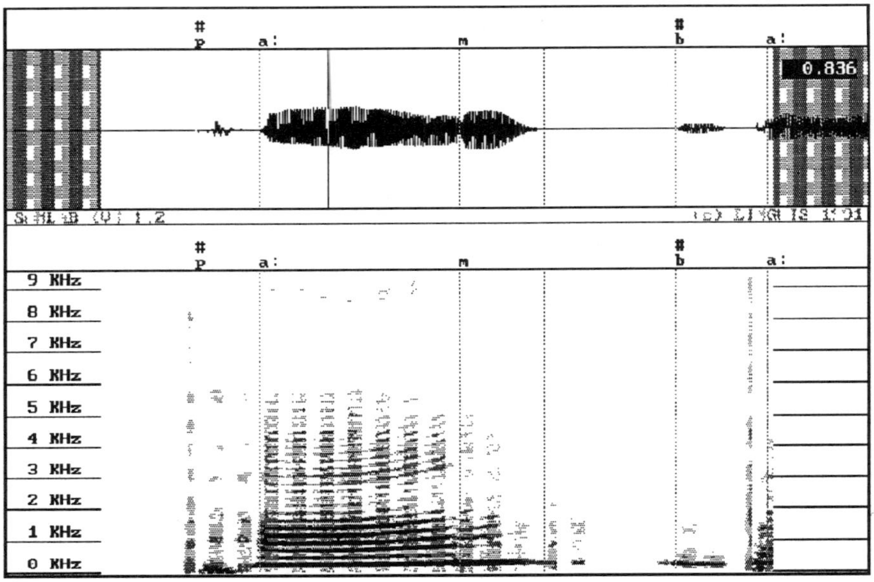

Figure 2.4: Signal annotation of the sequence [pɑːm]

annotations of the individual features were approximated on the basis of the signal annotations and the resulting structure is given in Figure 2.5.

The features *occlusion*, *transient* and *noise* define a more complex feature *plosive*. These features are interpreted as phonetic events which compositionally define phonological events. In the course of this book, different types of events are introduced ranging from simplex events such as these to more complex structures representing syllable or coda events. The simplex and the complex can be related within a declarative framework. Phonetic organisation is considered to play a similar role to phonological organisation in that constraints are placed on event relations describing possible descriptions. In Chapter 5, an event-based autosegmental phonotactics (set of well-formedness constraints) for German is presented in which possible syllable structures are defined in terms of overlap and precedence relations between event types (as opposed to event tokens). A *primary tier*, is defined which relates all the tiers. A primary tier which corresponds to the timing (CV-) tier of Clements & Keyser (1983), defines a perspective for looking at other tiers. In the case of the

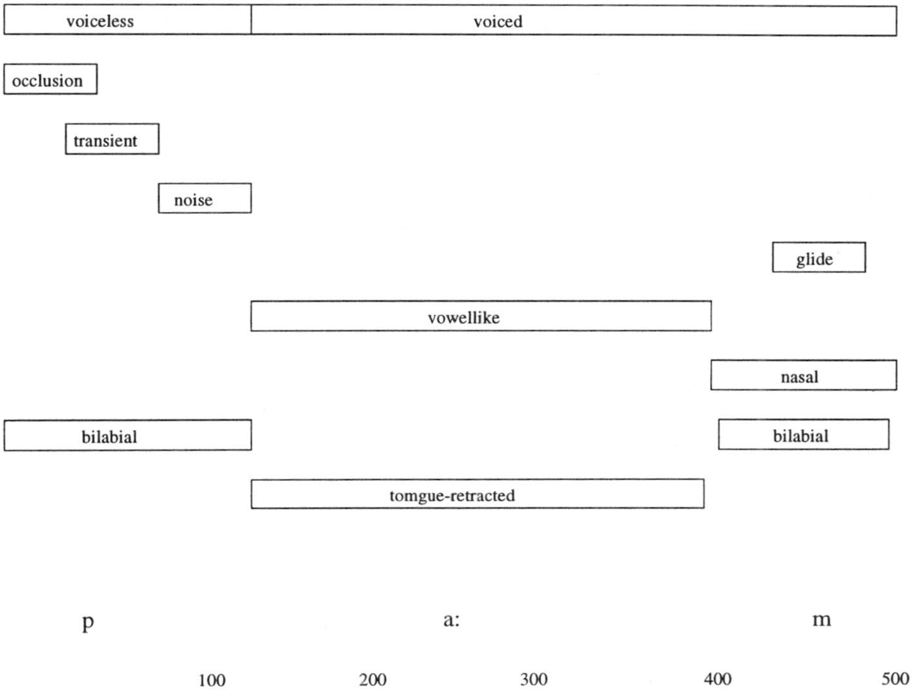

Figure 2.5: Temporal feature structure for the sequence [pɑːm]

phonotactics, a primary tier has been chosen which describes the canonical segments which constitute the syllable structure. However, it is possible to visualise other primary tiers which describe syllables from another perspective (e.g. sonority, voicing, etc.). In Chapter 5, it is shown that it is possible to define tactics (well-formedness constraints) for other phonological events (e.g. a plosive event) analogously to the syllable event phonotactics. Such phonological events also have their own primary tiers. The primary tier provides a basis for interpretation within an event-based framework.

Time Map Phonology is an event-based autosegmental approach to spoken language processing which is motivated by the recent developments in declarative phonology discussed at the beginning of the chapter. This approach combines attributes of autosegmental and articulatory phonology, and adopts many features of the event logic which provides the basis of constraint-based phonology (Bird, 1990). In particular, the modelling of traditional phonological processes such as elision, epenthesis or assimilation is analogous to the articulatory phonology approach in terms of degree of overlap (gestural magnitude and intergestural overlap) rather than deletion or insertion, is appropriate for spoken language recognition. The notion of autonomous features is appealing for both phonological description (e.g. for prosodic phenomena) and for phonetic description since overlapping information is permitted. The event concept also provides the means for describing other phenomena which are usually considered problematic in phonology. Ambisyllabicity (cf. Kahn, 1976), for example, is expressed elegantly in this theory as an overlap of syllable-coda and syllable-onset events.

In this approach, events of different kinds are assumed to play a role at all levels of linguistic description (e.g. occlusion events, plosives events, syllable-onset events, syllable events, morph events,..., noun-phrase events etc.). *Time Map Phonology* is concerned with the extra-lexical phonological domain of description; that is to say, it makes no reference to morphological information. This domain is sometimes referred to as the post-lexical domain; this term has procedural connotations and for this reason the term extra-lexical is preferred. The somewhat controversial claim is made in this book that this domain is 'phonology proper'. It was mentioned above, that in many cases it is unclear which domain a theory is intended to describe. Linear generative phonology describes both morphophonological and phonological phenomena and most theories since have assumed that this is the domain which must be adequately described by a theory of phonology. The lexical domain as described by Bleiching (1991), Gibbon (1992a), Gibbon & Langer (1992) and Pampel (1992), is closely related to the work described here but it is considered to be a separate domain.

Now that the empirical domain P of *Time Map Phonology* has been defined as the extra-lexical domain, the discussion turns to the adequacy of this approach for computational modelling. The following specific claims are made:

Empirical Adequacy: *Time Map Phonology* is at least descriptively adequate for the empirical domain P. It provides the ability to generalise beyond *actual* phonological objects in the domain P, making predictions about *potential* phonological objects which are well-formed in P. In this way, a distinction is made between accidental and systematic gaps of P. Descriptive adequacy is discussed in depth in connection with syllable phonotactics in Chapter 5.

Representational Adequacy: The formalism used in *Time Map Phonology* is representationally adequate in that it is capable of expressing all and only well-formed phonological objects of the domain P in a notation appropriate to P which has a formal syntax and semantics. These points are demonstrated in the remaining chapters of the book. Chapter 3 is concerned with expressive power required for the domain. Chapter 4 introduces the primitives, axioms and rules of inference for *Time Map Phonology* and Chapter 5 presents a notation which restricts the formalism to the description of regular expressions. The theory is therefore empirically testable.

This approach fulfils the declarative adequacy criteria for computational phonological modelling. Procedural adequacy of the computational model has also been considered in an application in spoken language recognition. In Chapter 7, an operational semantics for this model is presented together with a concrete implementation within the linguistic word recognition component of a speech recognition system; procedural adequacy is then measured in statistical terms on this basis in Chapter 9.

3

FINITE STATE TECHNIQUES IN COMPUTATIONAL PHONOLOGY

Finite state phonology has contributed substantially to the declarative paradigm. It has provided a formal basis for arguments in favour of a non-derivational phonology. It would be beyond the scope of this book to give a review of all developments in this area. This chapter aims to give an overview of the main developments in finite state phonology and discuss the role they can play in spoken language processing. In Section 3.1 the formal stage is set. Section 3.2 presents finite state approaches to segmental phonology and Section 3.3 introduces more recent work in connection with so-called nonlinear or perhaps more appropriately in this context *multi-linear* phonology. Section 3.4 concludes the chapter with a discussion of the implications of finite state phonology for phonological parsing in a spoken language system.

3.1 SOME DEFINITIONS

This section is primarily concerned with those formal definitions from automata theory which are necessary for an understanding of finite state phonology. The definitions are based on those given by Aho & Ullman (1972), Hopcroft & Ullman (1979) and Partee et al. (1990). The basic device of finite state phonology is the finite state automaton (FSA). Partee et al (1990:455) provide a simple informal definition of a finite state automaton as "an abstract computing device that receives a string of symbols as input, reads this string one symbol at a time from left to right, and after reading the last symbol halts and signifies either acceptance or rejection of the input". The input is usually regarded in terms of a single tape which is scanned by the machine.

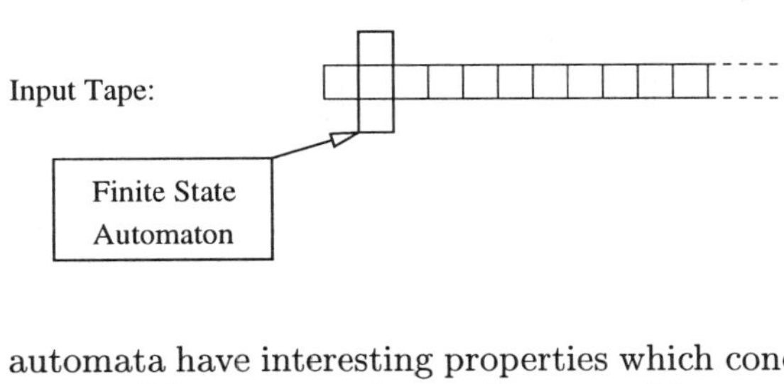

Finite state automata have interesting properties which concern the type of languages they accept. These properties are just stated here, no formal proofs are given (but cf. Aho & Ullman (1972), for example). All finite sets of strings are accepted by finite state automata. This turns out to be a very useful property for phonological description. Furthermore the languages defined by finite state automata are exactly the Type 3 languages in the Chomsky Hierarchy, otherwise known as the regular languages.

A finite state automaton accepts all regular languages. A distinction is made between deterministic finite state automata and nondeterministic finite state automata. Determinism refers to the choices which an automaton has in order to make the next move. A finite state automaton is said to be deterministic if, during recognition, its behaviour is completely determined by the state it is in and the symbol it is looking at. In other words, the automaton has no choice; there is only one possible next state. The nondeterministic finite automaton, on the other hand, may have several possible next moves. It is known that for every non-deterministic finite state automaton (NDFSA) there exists a deterministic finite state automaton (DFSA) which accepts the same language. The deterministic finite state automaton is defined formally as follows:

Definition 3.1 *A deterministic finite state automaton is a quintuple*

$$DFSA = <Q, \sum, \delta, q_0, F>$$

where

- Q *: a finite, nonempty set of states*
- $\sum$ *: a finite input alphabet*
- δ *: a function mapping from* $Q \times \sum$ *into* Q
- $q_0 \in Q$ *: the initial state*
- $F \subseteq Q$ *: the set of final states*

The work of an automaton can be best understood in an example. The transition function of a DFSA can be represented graphically as a transition diagram. An example is provided in Figure 3.1. The set of states Q of this diagram is

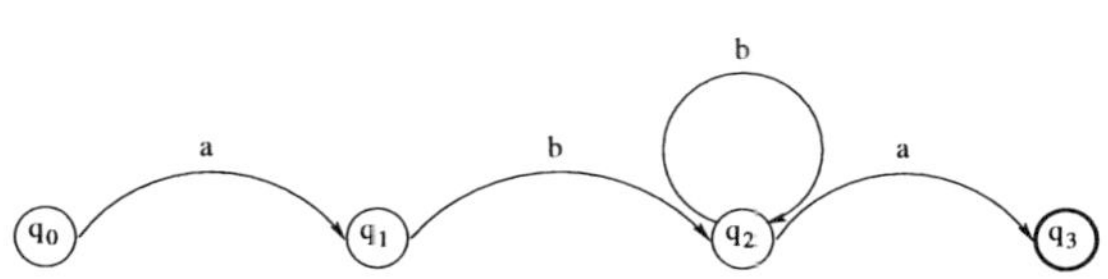

Figure 3.1: Transition diagram

$\{q_0, q_1, q_2, q_3\}$ where q_0 is the initial state and q_3 is the final state. The input alphabet $\sum$ is {a,b} and the δ function is defined as follows:

$$\delta(q_0,a) = q_1$$
$$\delta(q_1,b) = q_2$$
$$\delta(q_2,b) = q_2$$
$$\delta(q_2,a) = q_3$$

The automaton accepts the regular language ab^+a (i.e. abb*a). The $^+$ is the Kleene $^+$ which means 1 to n iterations rather than * which means 0 to n. This function may also be represented as a matrix.

	a	b
q_0	q_1	
q_1		q_2
q_2	q_3	q_2
q_3		

Given an input string *abbbba* as a tape in Figure 3.1, the automaton moves from q_0 to q_1 accepting the symbol *a*, then from q_1 to q_2 accepting *b*, loops from q_2 to q_2 accepting *b* a further three times and then accepts *a* by moving to q_3 which is the final state. The input string *aabba* would be rejected, however since the second symbol must always be a *b*. This is the only example of this kind which is given in this chapter in order to illustrate the basic device. Further definitions are given below without such examples.

The behaviour of a nondeterministic finite state automaton is not completely determined by the state it is in and the symbol it is looking at. It has more than one possible next states. It is defined formally as follows:

Definition 3.2 *A nondeterministic finite state automaton is a quintuple*

$$NDFSA = <Q, \sum, \delta, q_0, F>$$

where

- Q *: a finite, nonempty set of states*
- $\sum$ *: a finite input alphabet*
- δ *: a function mapping from* $Q \times \sum$ *into* $\wp(Q)$
- $q_0 \in Q$ *: the initial state*
- $F \subseteq Q$ *: the set of final states*

The nondeterminism is expressed in the δ function which maps into the power set of Q rather than into single elements as in the deterministic case.

The definitions above referred to single tape input. It is also possible to have a finite machine which refers to two tapes. Such a machine is known as a finite state transducer. Generally, one tape is designated as the input tape

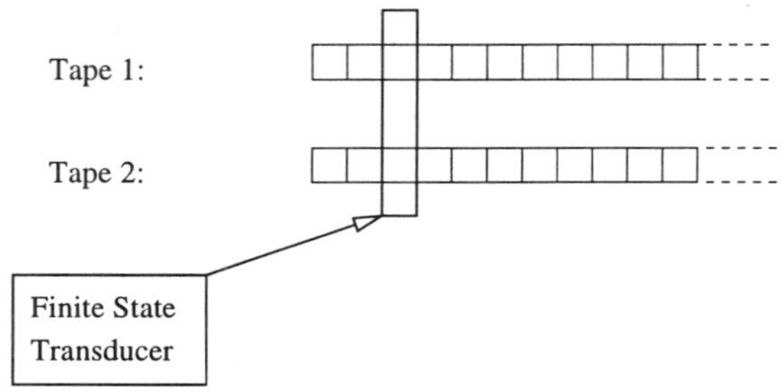

and the other as the output tape. However,it may also be regarded as an FSA where the elements of $\sum$ are pairs, one member of each pair for each tape. A finite state transducer is also distinguished with respect to determinism. The deterministic finite state transducer (DFST) is defined formally as follows:

Definition 3.3 *A deterministic finite state transducer is a sextuple*

$$DFST = <Q, \sum, \Delta, \delta, q_0, F>$$

where

- Q *: a finite, nonempty set of states*
- $\sum$ *: a finite input alphabet*
- Δ *: a finite output alphabet*
- δ *: a function mapping from* $Q \times \sum$ *into* $Q \times \Delta*$
- $q_0 \in Q$ *: the initial state*
- $F \subseteq Q$*:the set of final states*

This is the definition given in Aho & Ullman (1972:224). Here only a single function δ is defined. The δ function scans a tape with symbols from $\sum$ and a tape with symbols from Δ. It is also possible to separate this into two functions δ and λ where the former is a function mapping from Q $\times$ $\sum$ into Q and the latter is a function mapping from Q $\times$ Δ into Q. As mentioned above, these are usually regarded as input and output tapes but this interpretation is not obligatory. They may also be interpreted as two input tapes. The convention for representing symbols on corresponding tapes is $\alpha : \beta$ where $\alpha \in \sum$ and $\beta \in \Delta$. A nondeterministic finite state transducer is defined analogously to its deterministic counterpart except that the δ function maps into finite subsets of states as with the nondeterministic automaton.

Definition 3.4 *A nondeterministic finite state transducer is a sextuple*

$$NDFST = <Q, \sum, \Delta, \delta, q_0, F>$$

where

- Q *: a finite, nonempty set of states*
- $\sum$ *: a finite input alphabet*
- Δ *: a finite output alphabet*
- δ *: function mapping from* $Q \times \sum$ *into finite subsets of* $Q \times \Delta^*$
- $q_0 \in Q$ *: the initial state*
- $F \subseteq Q$*:the set of final states*

The devices which provide the basis for finite state phonology have now been defined. In the next two sections, it is shown how these devices are applied, together with certain operations, initially in linear phonological description and then with respect to multi-linear autosegmental representations. Rather than defining all possible operations on finite automata and transducers formally in this chapter, only those operations which are relevant for the individual approaches to finite state phonology are given in the relevant sections.

3.2 FINITE STATE PHONOLOGY: LINEAR REPRESENTATIONS

The origins of finite state phonology are usually traced back to the work of Johnson (1972). Although simple finite state models of linguistic description have existed since the 1950s (e.g. Shannon & Weaver, 1949; Hockett, 1955), it was the theory of linear generative phonology (Chomsky & Halle, 1968) which gave rise to discussions of restrictedness and later to nonproceduralality (cf. the discussion in Chapter 2). Johnson claims that the rewrite rule formalism of generative phonology is not appropriate to the restricted types of mappings required between phonological strings. He develops his arguments in favour of this claim by reviewing the schematic notation presented in Chapter 10 of Chomsky & Halle (1968:330-399) and showing that such schemata can be represented by all and only those sets which represent regular expressions (Johnson, 1972:22f.). Since Chomsky & Halle propose that such a notation is adequate for phonological descriptions, Johnson concludes that phonology, in contradiction to syntax, can be described in terms of a regular (Type 3) language. The unrestricted rewrite rule formalism, on the other hand, has the expressive power of a Turing machine and therefore allowing phonological rules to simulate such rewriting systems does not indicate the restricted nature of the mapping such rules are required to perform (Johnson, 1972:43).

The aim of Johnson's work is to show that phonological rules can be restricted to those which can be modelled as finite state transducers which map from regular expression to regular expression. First, he shows that in the case of simultaneous rule application (cf. Chapter 2), phonological rules can be expressed by finite state transducers since it is the original string that determines the application of the rule to any point in the string. Iterative application, on the other hand, does not, as it stands, impose the required finite state restriction since the original string does not always determine the application of a later rule; one phonological rule may change the context for application of following rules. Johnson introduces another type of rule, known as the linear rule, which has some of the properties of the iterative type. By restricting the application of a linear rule to strictly left-right application or strictly right-left application, the power of iterative rule application can be reduced to finite state (Johnson, 1972:61). This restriction means that no rule can be reapplied to its own output. After discussing many examples, Johnson concludes his thesis with the claim that only linear rules are necessary for phonology. Although simultaneous rules can also be represented by finite state transducers, he is unable to find any examples of simultaneous rule application which cannot be modelled by the linear type (1972:79). A more detailed summary of Johnson's proposals (and also other proposals which are mentioned below) can be found in Ritchie et al. (1992:16ff).

Although finite state automata did find some applications within the area of intonation ('t Hart & Collier, 1975; Pierrehumbert, 1980), it was not until 1981 that the notion of using finite state transducers for representing generative phonological rules was taken seriously. At a 1981 conference, Kaplan & Kay presented results of work on representing the phonological rules of generative phonology by sequential finite state transducers. A paper describing the mathematical foundations of their proposals was published after many years in 1994 (Kaplan & Kay, 1994).

Kaplan and Kay demonstrate that by imposing the restriction on all phonological rules that they may not apply to their own output, a large class of rewriting rules can be mimicked by finite state transducers. This constraint refers to symbols in the representation; once a symbol has been changed by a rule, it may not be changed again by another rule. This is equivalent to a combination of the left-linear application and the right-linear application of Johnson (1972). This restriction constrains the mapping, as also noted by Johnson, to a regular relation; a mapping between phonological strings as regular expressions. For each phonological rule, Kaplan and Kay propose a separate finite state transducer and in order to get from an underlying to a surface form, the finite state transducers are applied sequentially. Thus the output tape of one transducer is the input tape of the next transducer. By employing the notion of composition of transducers, they show that the sequentially applied transducers can be compiled into (or encoded as) a single transducer which maps the underlying form directly onto the surface form. Composition of finite state transducers can be understood analogously to the composition of functions. Given a transducer T_1 which maps RE_1 onto RE_2 and a transducer T_2 which maps RE_2 onto RE_3, the transducer T_3 is the composition of T_1 and T_2 which maps RE_1 onto RE_3. Composition can be applied repeatedly to a sequence of transducers until only a single transducer remains. A diagrammatic representation is given in Figure 3.2.

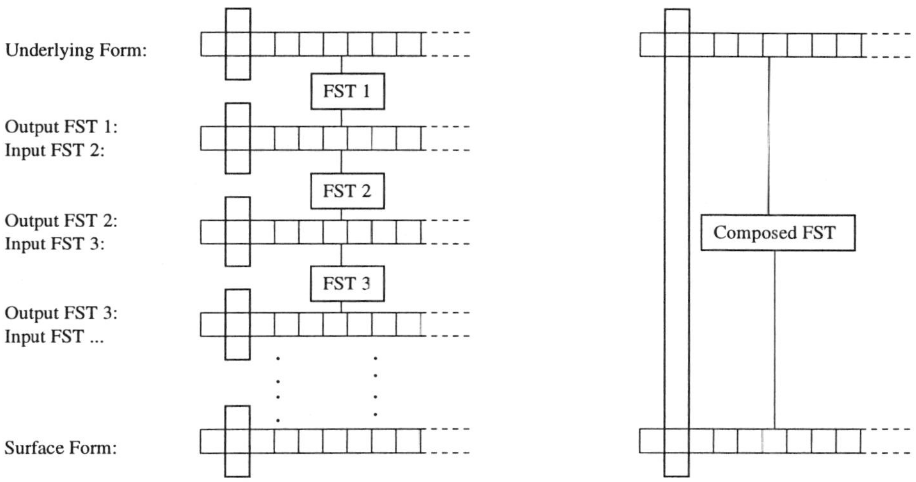

Figure 3.2: Diagrammatic representation of composed FST

An example of the composition of two transducers describing typical rules of generative phonology can be found in Karttunen (1991). Figures 3.3 and 3.4 show the transducers representing the rules 3.1 and 3.2 (note that x:x refers to all pairs of symbols which are not explicitly otherwise mentioned in the transducer).

$$N \rightarrow m/\_p \tag{3.1}$$

$$p \rightarrow m/m\_ \tag{3.2}$$

The composed finite state transducer for the two rules can be computed by defining new states and arc descriptions based on each of the individual transducers as described informally here:

1. an initial state (00) corresponding to the initial states of the individual transducers,

2. final states (00, 10 and 01) based on the final states of the individual transducers and

3. a new set of states and arc descriptions such that for any arc a new description x:z is assumed whereby x:y is on the arc of one transducer and y:z is on the arc of the second transducer.

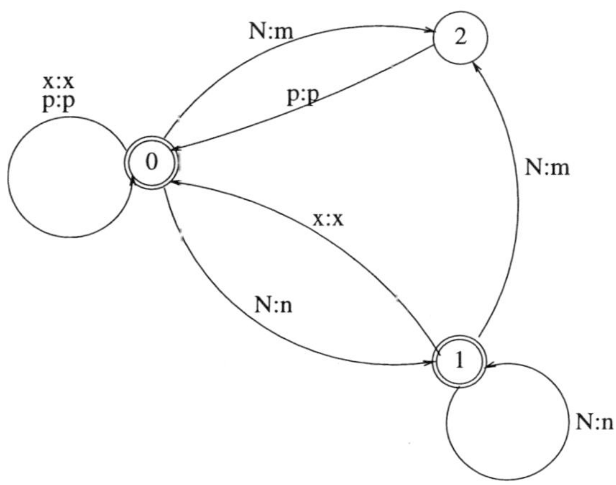

Figure 3.3: FST for Rule 3.1

The δ functions for the individual transducers (the x represents all phonemes which are not specifically named) are given in Tables 3.1 and 3.2.

Table 3.1: Transducer for Rule 3.1

Q : {0, 1, 2}					
q_0 : 0					
F : {0, 1}					
$\sum$: {p, N, x}					
Δ : {p, m, n, x}					
(1.1)	δ(	0,	p:p,	0	)
(1.2)	δ(	0,	x:x,	0	)
(1.3)	δ(	0,	N:n,	1	)
(1.4)	δ(	0,	N:m,	2	)
(1.5)	δ(	1,	N:n,	1	)
(1.6)	δ(	1,	N:m,	2	)
(1.7)	δ(	1,	x:x,	0	)
(1.8)	δ(	2,	p:p,	0	)

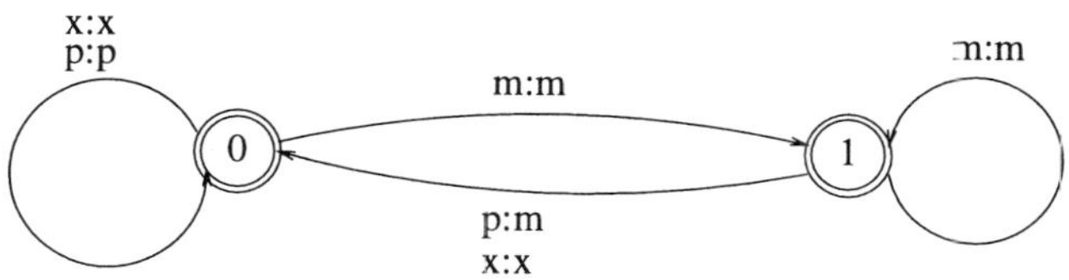

Figure 3.4: FST for Rule 3.2

Table 3.2: Transducer for Rule 3.2

Q : {0, 1}					
q_0 : 0					
F : {0, 1}					
$\sum$: {p, m, x}					
Δ : {p, m, x}					
(2.1)	δ(	0,	p:p,	0	)
(2.2)	δ(	0,	x:x,	0	)
(2.3)	δ(	0,	m:m,	1	)
(2.4)	δ(	1,	m:m,	1	)
(2.5)	δ(	1,	p:m,	0	)
(2.6)	δ(	1,	x:x,	0	)

Composing these transducers yields the transducer given in Table 3.3. Since state 20 and 11 are not initial states and no arc leads to them (3.10), (3.14), (3.15) and (3.16) can be removed giving the transducer shown in Figure 3.5.

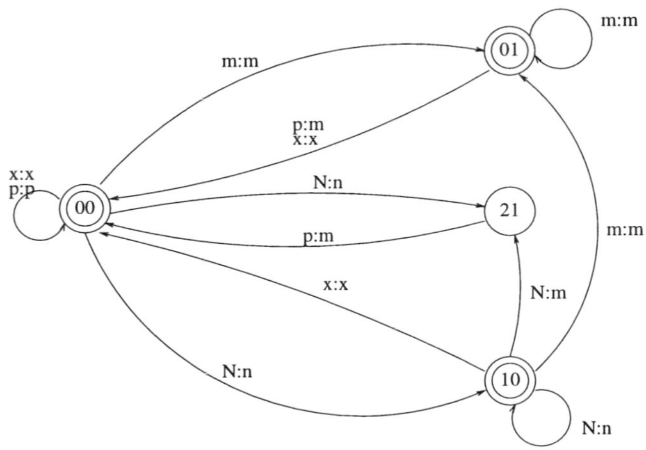

Figure 3.5: Composed FST for Rules 3.1 and 3.2

Table 3.3: Composed Transducer for Rules 3.1 and 3.2

Q : {00, 10, *20*, 01, *11*, 21}
q_0 : 00
F : {00, 10, 01}
$\sum$: {p, m, N, x}
Δ : {p, m, n, x}

(3.1)	δ(00, p:p, 00)	(1.1 and 2.1)
(3.2)	δ(00, x:x, 00)	(1.2 and 2.2)
(3.3)	δ(00, m:m, 01)	(1.2 and 2.3) since m:m in second transducer must be made explicit
(3.4)	δ(00, N:n, 10)	(1.3 and 2.2) since N:n in second transducer is x:x
(3.5)	δ(00, N:m, 21)	(1.4 and 2.3)
(3.6)	δ(10, N:n, 10)	(1.5 and 2.2) since N:n in second transducer is x:x
(3.7)	δ(10, x:x, 00)	(1.6 and 2.2)
(3.8)	δ(10, m:m, 01)	(1.6 and 2.4) since m:m in second transducer must be made explicit
(3.9)	δ(10, N:m, 21)	(1.7 and 2.4)
(3.10)	δ(20, p:p, 00)	(1.8 and 2.1)
(3.11)	δ(01, p:m, 00)	(1.1 and 2.5)
(3.12)	δ(01, x:x, 00)	(1.2 and 2.6)
(3.13)	δ(01, m:m, 01)	(1.2 and 2.4)
(3.14)	δ(11, N:n, 10)	(1.5 and 2.6) since N:n in second transducer is x:x
(3.15)	δ(11, x:x, 00)	(1.6 and 2.6)
(3.16)	δ(11, m:m, 01)	(1.6 and 2.4) since m:m in second transducer must be made explicit
(3.17)	δ(21, p:m, 00)	(1.7 and 2.5)

The advantage of the Kaplan/Kay discovery is that the notion of rule ordering in generative phonology (cf. Chapter 2) loses its impact. Phenomena which, according to generative phonologists, could only be described by a set of ordered rules, were given a description in terms of composable finite state transducers. Phonological rule ordering is accompanied by a problem of reversibility, namely the difficulty of trying to find an underlying representation from a surface form. The finite transducer representation does not have this problem since the mapping applied directly between two levels; given one level it is possible to derive the other directly. The intermediate levels of representation, which have no cognitive meaning, are no longer necessary; this point is discussed further below. The finite state transducers have the advantage of formal and computational simplicity. A disadvantage of the original proposal made by Kaplan & Kay (1981) was that for practical purposes the encoding in terms of a single finite state transducer becomes too complex. In their 1994 paper, however, they state that, for the domain of phonological and orthographic systems, they have considered their formalism from a more abstract viewpoint, developing an appropriate set of high level algebraic operators for reasoning about languages and relations and for justifying a corresponding set of operators and automata for computation (Kaplan & Kay, 1994: 336). They demonstrate in detail how the rules of generative phonology can be restated as a single finite state transducer using these operations. Further algorithms for parallel intersection and serial composition of finite state transducers can be found in Reape & Thompson (1988).

In 1983, Koskenniemi proposed a two-level formalism designed to overcome the above mentioned disadvantage of Kaplan & Kays' single composed finite state transducer. Koskenniemi describes his model as morphological, although he also takes orthography and phonology into account. It consists of a lexicon system and a set of two-level rules. The lexicon lists the morphemes of the language and describes the morphotactics in terms of a finite state automaton. The two-level rules relate the lexical and the surface representations. In this model there is no conception of an intermediate representation between the lexical and surface levels as with the Kaplan & Kay system. The phonological rules are compiled directly as finite automata or transducers and all rules act in parallel as filters on the well-formedness of the representations; they restrict the possible correspondences between the levels. The two-level rules avoid problematic rule interactions and are conceptually simpler than generative rewrite rules (Koskenniemi, 1983b:9). For the discussion which follows, the two-level rules are considered independently of the lexicon concept which is used.

Two-level rules declaratively specify constraints on the well-formedness of lexical and surface representations. Together they specify a regular relation. The following types of two-level rule are introduced (Koskenniemi, 1983b:36f); Koskenniemi's symbols have been replaced by Greek letters:

Context restriction rule: $\alpha : \beta \Rightarrow / \gamma \_ \delta$ which is interpreted as α is realised as β only in the context $\gamma \_ \delta$.

Surface coercion rule: $\alpha : \beta \Leftarrow / \gamma \_ \delta$ which is interpreted as α is always realised as β in the context $\gamma \_ \delta$.

Composite rule: $\alpha : \beta \Leftrightarrow / \gamma \_ \delta$ which is interpreted as α is realised as β in the context $\gamma \_ \delta$ and nowhere else.

The $\alpha : \beta$ are termed a character pair. Using these rule types, finite state transducers for all possible lexical and surface forms are compiled. Koskenniemi uses special symbols for defining subsets of his alphabet such as **V** for all vowels, except those specified explicitly, and other symbols to trigger particular endings. A two-level rule of Finnish morphophonology in Koskenniemi's formalism would be the following:

$$\begin{matrix} i \\ j \end{matrix} \Leftrightarrow V + \_ \, V$$

This rule says that an i in the lexical representation is always realised as a j in the surface representation if the left context is a vowel V and a boundary symbol + and the right context is a vowel V and in no other context. This rule would be represented as a finite state transducer with the following δ function given by Koskenniemi (1983a:684) as a matrix:

	V V	+ 0	i i	i j	= 0	= =
1:	2	1	2	0	1	1
2:	2	3	2	0	2	1
3:	2	1	4	5	3	1
4.	0	4	0	0	4	1
5.	2	5	2	0	5	0

Koskenniemi refers to his two-level rules as automata rather than transducers. This is because the two levels always correspond as pairs and therefore the input tape of the automata can be regarded as a pair of symbols rather than symbols on two separate tapes. The transducer for the two-level rule is the following:

- Q : {1,2,3,4,5}
- $\sum$: {V,+,i,j,=}
- Δ : {V,0,i,j,=}

- δ : as in the matrix
- q_0 : 1
- F : {1,2,3}

If the δ function returns 0 as a state, this means that the mapping is undefined. A 0 as part of a character pair means that it is present in the lexical representation but is not seen at the surface level. The = symbol is a catch-all which means any character pair which has not explictly been mentioned. This can be specified in detail in terms of all other characters in the alphabet and therefore it is still regular. All two-level rules apply in parallel and a particular configuration is considered acceptable if all rules allow it through. This is shown in Figure 3.6.

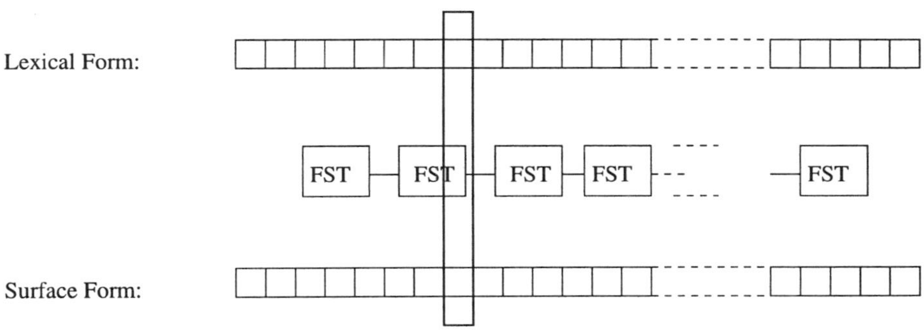

Figure 3.6: Diagrammatic representation of parallel finite state transducers

Koskenniemi applied his model to Finnish and in his thesis, he gives a complete description of Finnish phonology and morphology. He emphasises that the model is not language-specific. Indeed this work has provided the basis for much further work on the phonology and morphology of many different languages; the collection of articles in Texas Linguistic Forum 22 present applications to English, Japanese, and French. Other work in the area of segmental finite state phonology/morphology includes: Bear (1988, 1990), Görz & Paulus (1988) and Trost (1990).

Koskenniemi's work came in for some criticism at the time. Gazdar (1985) criticised the two-level model for using monadic segments instead of the feature systems of phonology and also pointed out that arbitrarily recursive patterning such as stress cannot be catered for by such a model. Despite these criticisms, the two-level model has played an important role in the development of finite state phonology and morphology since 1985. Ritchie et al. (1992:182) discuss the possibility of incorporating distinctive features and underspecified feature structures into the two-level model. They cite Carson (1988) in connection with proposals for manipulating features in this type of formalism.

In Chapter 5 of this book, a finite state model for German syllable structure is presented, based on the feature-based model described in Carson (1988). The motivation for such a model is provided by the discussion above. The set of German syllables is known to be finite and therefore the phonotactics can be represented by a non-cyclic finite state automaton. This model is fundamentally different from that of Koskenniemi in that it uses underspecified feature bundles rather than monadic segments which are equivalent to fully specified feature bundles. In Chapter 6, this model is extended to a transducer which maps between an allophonic and a phonological level. The transducer provides constraints on well-formedness of syllable structure at both these levels (cf. Carson-Berndsen et al. (1989)). A further extension to the transducer model allows speech variant rules to be represented in the same manner as the metarules described by Kay (1983). These metarules can then be incorporated into the transducer to form a new description which also accounts for variants. The syllable finite state transducer is used for phonological parsing in a segmental system in a manner similar to that described in Church (1987b, cf. the description in Chapter 1). The model proposed in Carson (1988), described in detail below, adheres to the finite state model using operations defined on the input and output tapes of the transducer in order to discard allophonic information as part of the parsing process rather than having a separate canonicalisation step. The transducer maps allophonic syllable structures onto phonological syllable structures. Being a two-tape model, it is directionally neutral and thus can map also from the phonological level to the allophonic level, although with a lesser degree of determinism. Here, as in the two-level model, no intermediate representation between the levels is assumed. However, contrary to the two-level model, it does not need to cope with phenomena which require a hierarchical structure and thus remains within a linear domain.

The phonological parser of Church (1987b) does not explicitly use a finite state automaton formulation of the rules of syllable structure. Instead, Church proposes a system of phrase structure rules which contain allophonic information to represent syllable structure. Although the phrase structure rule formalism is not restricted to describing regular expressions, Church claims that his grammar is only finite state (1987b:110) and indeed, in the examples he provides, the local hierarchies defined by the phrase structure rules are right-branching and therefore regular.

To close the discussion of segmental finite state phonology, the notion of declarativity is revisited. Due to the fact that finite state phonology and morphology has made reference to automata and transducers, it has been regarded as a procedural interpretation of linguistic facts. While there is no doubt that such a procedural interpretation can be given in this way, this has not been the aim of finite state phonology and morphology, however, as should be clear from the discussion above. Finite state transducers are simply used as one way of representing regular relations, that is to say, relations between regular languages. Strangely enough, for generative phonologists, finite state phonology is too declarative. Karttunen (1991) gives examples of how phenomena which are traditionally thought to require rule ordering can be described within the finite state formalism (cf. Chapter 2). He also shows that it is possible to state certain phonological facts more succinctly using finite state devices than if generative phonological rules were employed. Finally he cites further support in favour of the finite state cause in the proposals of Lakoff (1987) for a cognitive phonology. The proposals of Lakoff (1987) have been developed further by Wheeler & Touretzky (1989, 1990) who have implemented a *many maps* architecture for a connectionist implementation. They distinguish between three levels: the morphemic level (M), the phonemic level (P) and the phonetic level (F). The levels are represented by separate (transducer-like) tapes between which mappings are carried out using matrix operations to perform deletions, insertions and mutations. This corresponds to the cross-level constructions proposed by Lakoff. However, they state that it is unclear how to represent well-formedness constraints (intra-level constraints) in a connectionist framework (1989:1) and therefore they constrain their model by not allowing intra-level constraints (1990:17). However, recent developments in phonology (as described in Chapter 2) have shown the statement of constraints on the well-formedness of phonological descriptions to be of major importance. In addition, a model which aims at a complete phonological description of a language will require general constraints on well-formedness in order to distinguish between *actual* and *potential* forms.

3.3 FINITE-STATE PHONOLOGY: MULTI-LINEAR REPRESENTATIONS

As was seen in the previous section, much work has been done in the area of segmental finite state phonology. However, phonology is not restricted to purely segmental theories as was seen in Chapter 2. This section introduces proposals for multi-linear finite state phonology in line with the work in autosegmental phonology. Kay (1987) offered the first proposal to extend the finite state formalism to cater for autosegmental representation by using an n-tape finite state transducer. Using data from nonconcatenative Arabic verb morphology, Kay demonstrates how such a transducer would work. The transducer in the example has four tapes which means that the representation is in terms of a quadruple rather than the pairs of the two-level model. Kay introduces the following conventions which direct the movement of the transducer: a symbol in brackets [ ] means that the tape does not advance, a symbol in braces { } means that the tape moves if the symbol in question is the last one on the tape; all other symbols function as in the two-tape transducer. The alphabet for the four-tape transducer is a set of quadruples or *frames* which contain brackets, braces and conventional symbols. A possible configuration of the four-tape finite state transducer is presented graphically in Figure 3.7. It is assumed in this configuration that the brackets have applied somewhere already, that is to say, one tape has not moved on, although the others have.

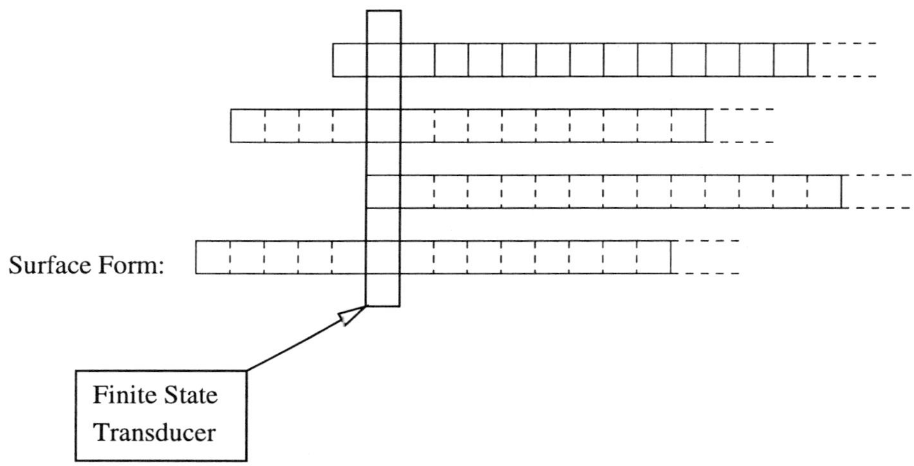

Figure 3.7: Diagrammatic representation of a configuration of a 4-tape transducer

In the rest of the paper, Kay gives a detailed account of how the four-tape model is applied to generate a surface form from the prosodic templates (or autosegmental tiers). He discusses phenomena which can be catered for by the model, for instance gemination, and concludes by discussing problematic issues. However, the problems discussed by Kay concern the data he is trying to describe. The tiers which are given do not seem suitable to describe all required forms with this model. An n-tape model allows for some of the information in the description to be separated onto more tiers than are provided. This should not be regarded as a defect of the model. Rather this insight should be taken as an indication that there is no intrinsic restriction of this sort on the number of tiers which can occur in an autosegmental description. This would seem to be the lesson which has been learned from Johnson (1972); it is perhaps more appropriate to change the description rather than to discard the formalism. Whether there are empirical constraints on the number and type of tiers is a distinct question.

The prosodic parser for parallel processing of both segmental and suprasegmental phenomena on autonomous tiers (Gibbon & Braun, 1988a) described in Chapter 1, differs from the Kay (1987) single multi-tape transducer for autosegmental relations in that it uses a configuration of several transducers within a parallel, sequential, incremental or parallel, hierarchical, incremental processing architecture. The individual transducers are formally similar to the structure of Koskenniemi's (1983b) two-level transducers (cf. Section 3.2) and to a lesser extent with Hidden Markov transducers. This approach uses finite state transducers to simulate feature modules such as acoustic edge detectors and contour detection, and parallel and cascaded transducers to simulate clusters of features and autosegmental tiers (1988a:202). Related work on the nonlinear finite state processing of tone languages can be found in Gibbon (1987).

Another approach to multi-linear finite state phonology is taken by Kornai (1991). In contrast to Kay, Kornai suggests a linear encoding of autosegmental representations. Surprisingly, Kornai does not quote the Kay (1987) approach. The primary concern of Kornai (1991) is the application of autosegmental phonology to a speech recognition task. His proposal is to provide a formalisation of autosegmental phonology which can serve as a basis for defining *structured Markov Models*. Such models are more appealing to speech engineers and he believes that in this way phonology can be incorporated more directly into the speech recognition process. Kornai criticises recent work on autosegmental phonology (e.g. Bird & Klein, 1990) which although it is sufficiently formal, does not take absolute (clock) time into account; a requirement which is needed in speech recognition. Furthermore, the phonetic interpretation of autosegmental representations has up to now disregarded the relationship between articulatory features and the acoustic signal (Kornai, 1991:11).

Kornai proposes a linearisation of autosegmental representations into a code which can be interpreted by a finite state device. The advantage of this method of autosegmental interpretation is that there is no need to add or delete association lines to a representation, as long as it is possible to check whether a given association line is present and if it is not to reject the representation (Kornai, 1991:64). The finite state device can be seen, as in the segmental case described in Section 3.2, to be a constraint on well-formed strings. Kornai defines an optimal code for autosegmental representations as having the following properties: it should be computable, invertible, iconic and compositional. However, he shows that the optimal code does not exist and therefore he relaxes the invertibility and iconicity properties.

Kornai presents two types of code: a *scanning code* which is relevant for association patterns, without requiring knowledge of the content of the tiers and a *triple code* which takes tier content into consideration. They are described below. In the initial version, the codes represent only two tiers (top tier and bottom tier) and they are linearised by a two-tape finite state transducer; Kornai uses the term *two-tape automaton*. Both codes are a record of the moves of the transducer and the association lines encountered during these moves. The presence of an association line in a representation is encoded as *1*, the absence of an association line is encoded as *0*. A move on the top tape of the transducer is encoded as *t* and a move on the bottom tape is encoded as *b*. The movement of the transducer is directed by the following advancement rules:

> If there are no further symbols on either tape, the machine stops. If there are no further symbols on one tape, the other tape is advanced by one. If there are no further association lines from x and y, both tapes move one step to the right, if there are further association lines from y, only the top tape moves, provided the move does not result in scanning G [the blank symbol]. (Kornai, 1991:27).

An example of how the scanning code for an autosegmental representation is generated is given in Figure 3.8. This example is based on Kornai (1991:29) for the triple code. Both types of code are illustrated here using the same example.

The triple code can be derived from the scanning code by 'flanking' these symbols with the corresponding tier symbols (Kornai, 1991:29). A first approximation to the triple code is the following linear representation: *a1Htk1Hc1Hd0Le1Mf0Hbf1Ltg1Lh0M*. This can be reduced by allowing for a symbol _ to represent spreading (association of an autosegment on one tier to more than one segment on another tier), resulting in *a1Hk1_c1H d0L e1M f0H_1L g1_h0M* as the triple code for the example in Figure 3.5. Kornai's own example of this triple code (1991:30) has a misprint: *h* should read as *H* in ...f0h_1L...

Autosegmental rules can be represented using the same linear encoding and finite control as for the representations. However, in order to avoid problems concerning the expressive power of autosegmental rules, Kornai prefers to devise automata which manipulate the autosegmental representations directly (1991:63). He proposes an *automatically advancing autosegmental automaton* (a^3-automaton) for analysis of autosegmental representations. The automaton is conventional in its definition except that the δ function defines the linearisation and interprets its input by advancing the tape by zero or one position (in line with the advancement rules above). Kornai also provides another type of automaton in the appendix which is not discussed further here.

Kornai emphasises that this encoding is only suitable for two-tiered representations. A generalisation of this encoding to multi-tiered representations gives rise to a vast amount of redundancy. A multi-tiered representation requires a different encoding and a different automaton for manipulating this different encoding. As a possible solution to this problem, Kornai suggests tier ordering (1992:72) with a *lazy advance* tape movement and unordered tiers

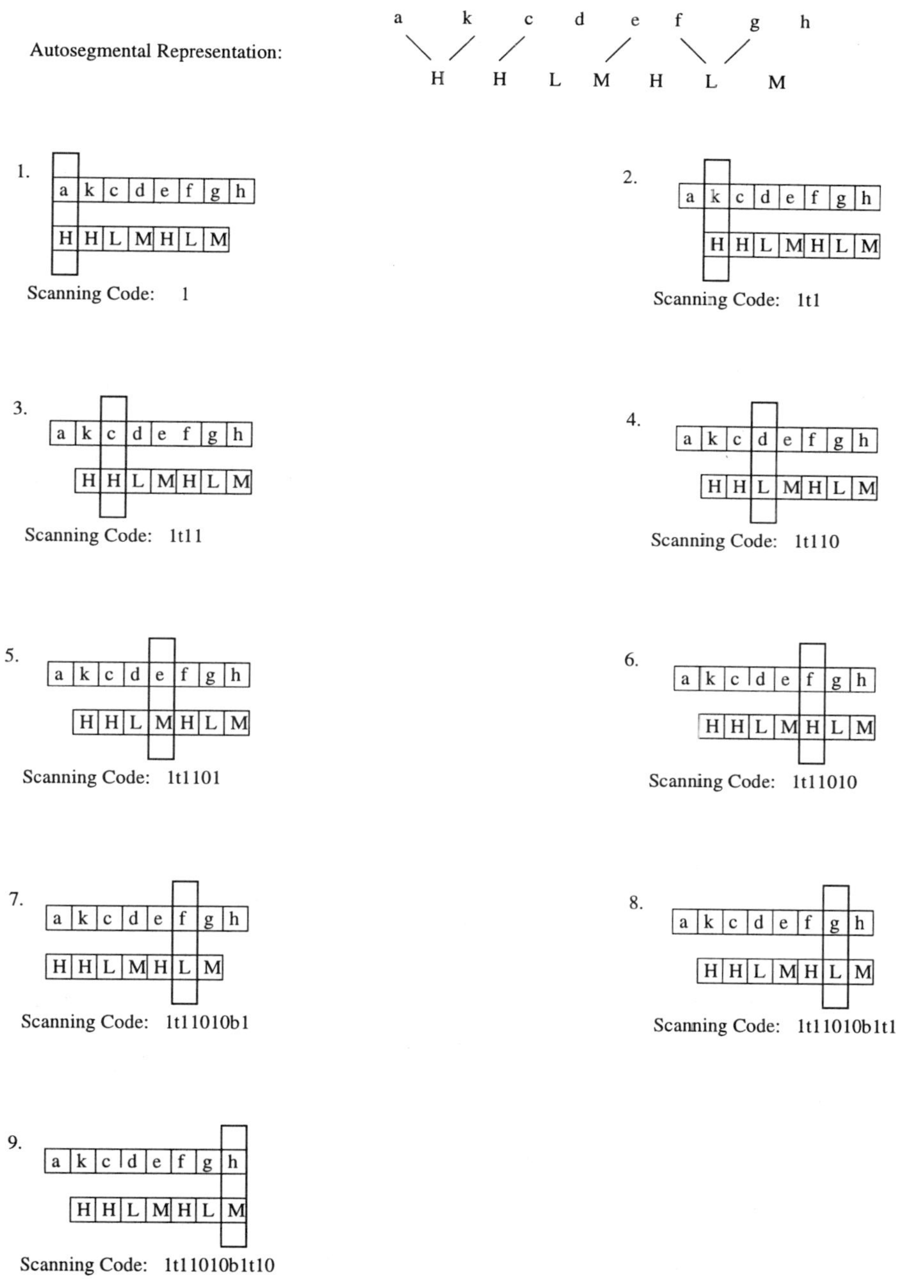

Figure 3.8: Construction of scanning code for an autosegmental representation

with an *eager advance* tape movement. These issues are not discussed any further by Kornai except that in this connection he also discusses the phonological notion of a distinguished timing tier which perhaps indicates that he prefers partially ordered tiers. In the rest of his thesis, Kornai deals with aspects of duration and synchronisation in autosegmental representations for the purposes of speech recognition and the development of the basis for the construction of a *structured Markov Model*. These matters are of no lesser importance that the encoding of autosegmental representations. However, since temporality is discussed in more detail in the next chapter, these aspects are not considered further. Kornai has not implemented his model.

A different approach to finite state interpretation of multi-linear representations is proposed in *Time Map Phonology* (cf. Carson-Berndsen (1991, 1992)) in which the redundancy problem is avoided. The types of autosegmental structures assumed by *Time Map Phonology* for the event-based phonological parsing of spoken German were introduced in Chapter 2. The proposal is presented briefly here; further details are provided in Chapters 4, 5 and 7. This approach combines a formal interpretation of autosegmental representations in terms of events with a computational interpretation of autosegmental representations in terms of a conventional finite state device as defined in Section 3.1. The phonological description of well-formedness for German, the event phonotactics, is a collection of constraints on overlap and immediate precedence relations between events. These constraints are defined with respect to a *primary tier* (timing tier in the sense of Clements & Keyser, 1983), in this case, the phonological segment tier which may be interpreted as an abstract timing tier. Phonological segments may also be regarded as events in the sense that they are structures which are constructed compositionally from other events. Each element of the primary tier defines (provides a reference to) constraints on overlap and immediate precedence of events. Precise definitions of events are provided in Chapter 4. The primary tier is represented as a finite state automaton which interprets these constraints. Transition diagrams of the primary tier automata for syllable events and other complex phonetic events are given Chapter 5. During analysis, the primary tier provides finite control and top-down constraints for the input event representation. This is illustrated in Figure 3.9. This finite state approach to the interpretation of autosegmental representations has many similarities with the n-tape and the linear encoding approach. However, it is more simple since it only requires movement of a single tape. Furthermore, the "encoding" of the primary tier is in terms of temporal relations on events, an encoding which can be interpreted by phonologists as well as by machines (this is not intended to be a criticism of Kornai's (1991) notation; however, it is somewhat difficult to read). Further details of this proposal are given in the chapters to come. The model has been implemented as part of a spoken language recognition system. This implementation is described in Chapter 7.

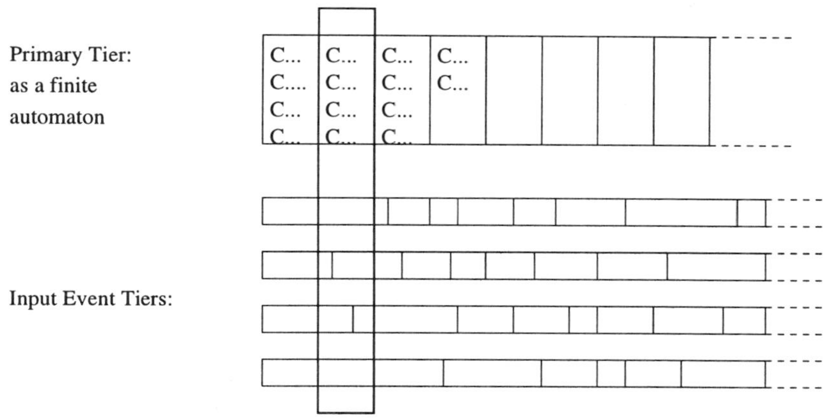

Figure 3.9: Diagrammatic representation of the primary tier automaton

Another approach to multi-linear finite state phonology is the one-level phonology proposed by Bird & Ellison (1992, subsequently published in 1994). They also propose the application of finite automata for the interpretation of autosegmental representations. However, their approach differs from those described above in that they use more unconventional state-labelled automata to represent individual tiers. Their device parallels the notion of a Moore machine. This is not a problem since it is equivalent to the conventional transition labelled automata (with output as a Mealy machine). However, they do not restrict their state-labelled automata to having a unique initial state. The automata have a different graphical representation as seen in Figure 3.10 which is based on those given in Bird & Ellison. The arrow heads represent initial states and the circled states represent final states when the automaton is allowed to terminate. This automaton represents a single autosegment on a tier.

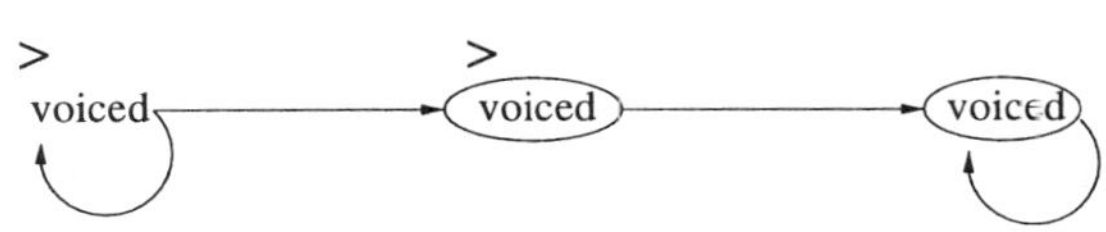

Figure 3.10: An example state labelled automaton

A state is allowed to map to itself. This loop allows the automaton to accept any string of one or more As. In order to represent association or overlap between autosegments of different tiers, the notion of synchronised automata is introduced. The synchronisation in Figure 3.11 represents an association line. It is clear to Bird & Ellison that such representations can become very complicated when dealing with more than one association and with more than two tiers. In order to combat this they show that by taking the product (or

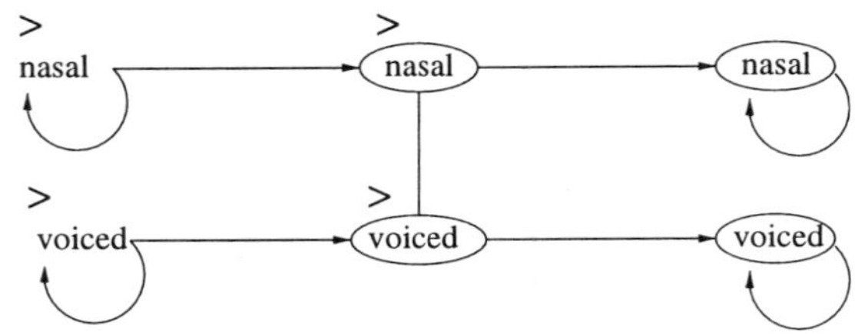

Figure 3.11: An example of a pinned automaton

intersection) of two state-labelled automata, a single simplified state-labelled automaton can be generated which interprets a multi-tiered autosegmental representation. This parallels the notion of the composed finite state transducer in Kaplan & Kay (1981). The resulting simplified finite automaton is very similar to the notion of the primary tier automaton described above. It is satisfying to see that although the Bird & Ellison proposal was developed independently of that presented in this book and Carson-Berndsen (1991, 1992), it draws similar conclusions regarding appropriate devices for interpreting multi-linear representations.

Bird & Ellison state "that a pair of adjacent intervals in which some property holds is indistinguishable from a *single* interval - the union of the first two intervals - during which that same property holds" (1994:65). It is for this reason also that loops are permitted in the automata. However, it is not clear why sequences of such segments with identical properties should be found in an input string since this notation suggests that a phonological segmentation has already been undertaken and the obligatory contour principle, the stipulation that no two adjacent segments have the same value, has in some sense been reversed. It would indicate that a 'sampling rate' is being assumed, whereby at some regular interval, feature detectors show which features are currently present. State looping is a standard technique for modelling this type of input (e.g. Hidden Markov Models, the finite state transducers in Gibbon & Braun, 1988a). However, Bird & Ellison (1994) do not discuss this input in any further detail.

Another approach to the computational modelling of autosegmental phonology has been proposed by Wiebe (1992). Wiebe proposes the use of *Multi-Tape Finite State Transducers* for modelling autosegmental phonology. These offer more computational power than traditional finite state transducers by allowing the read heads of the tapes to scan different parts of the representation at the same time which he claims is necessary for multilevel processing. He states that this is precisely what the Kay (1987) model requires but that this is not possible with a traditional finite state transducer. Wiebe also proposes a coding of autosegmental representations but in contrast to Kornai (1991), his coding remains nonlinear. The following is an example of his coding for the form *kattaba* 'he caused to write':

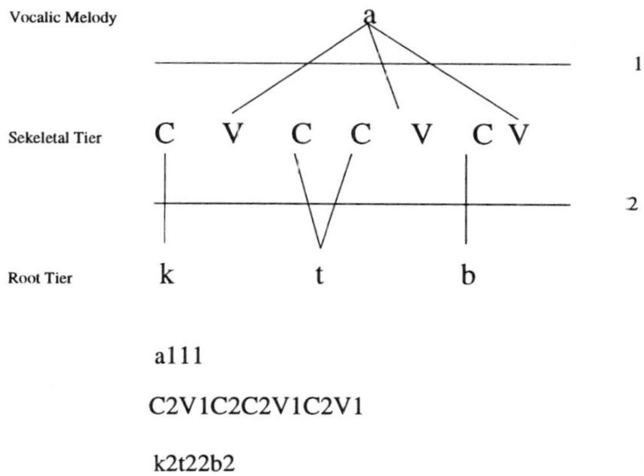

Figure 3.12: Wiebe's multilinear coding for autosegmental representations

A digit after a segment is used to represent an association line. Thus *C2* for example, indicates that there is an association line from C to a segment below level 2. The advantages of this encoding is that it has no problems handling more than two tiers and does not introduce the redundancy which results when Kornai's encoding is applied two more than two tiers.

Wiebe states explicitly (1992:120) that in contrast to the work of Bird & Ellison (1992), his primary aim in line with that of Kornai is to be faithful to the theory of autosegmental phonology as it stands (i.e. from a traditional generative viewpoint) rather than assuming a formal alternative as the basis for the computational model.

Since the model presented in this book was originally presented in 1993, Kirchhoff (1995b) has proposed a multilinear approach to speech recognition utilising six parallel arrangements of hidden Markov models combining stochastic signal modelling with finite state phonology. This approach employs phonetic features as the basic recognition units and syllables as the decision units for the application of delayed synchronisation. This approach differs from that proposed in this book in that synchronisation is not performed with respect to a primary tier but rather the alignment of features within the syllable domain remains temporally underspecified and synchronisation is accomplished at the syllable boundary points. In this way coarticulation phenomena are modelled. Preprocessing, training and feature recognition is carried out by HTK (Young & Woodland (1993)).

The tiers used (feature groups in Kirchhoff's terminology) are phonation, manner, place, front-back, centrality and roundness. Linguistic constraints on possible "feature-tactics" of these tiers within the syllable domain are extracted from the corpus during training and the resultant networks reflect the feature melodies found in the data. The phonation tier, for example, can be modelled by the following two networks (cf. Kirchhoff (1995b:80)):

1. <melody>:= +voi|-voi|sil
   <network>:= sil| <melody> |sil

2. <melody>:= +voi|+voi -voi|-voi +voi|-voi +voi -voi|-voi|sil
   <network>:= sil| <melody> |sil

Lexical mapping is performed by combining six parallel feature sequences with information from the automatic syllable boundary detector determining the synchronisation of the temporally underspecified syllable templates, and multi-level dynamic programming is used to find the optimal lexicon entry based on minimal distance.

In contrast to the multilinear models of Kay (1987), Kornai (1991), Bird & Ellison (1992) and Wiebe (1992), this approach has been operationalised and evaluated on spontaneous speech in scheduling tasks dialogues (28 hours training data and 3 hours 45 minutes test data) without a language model. The recognition results are described in detail in Kirchhoff (1996).

This concludes the review of multi-linear finite state phonology. Three of the approaches discussed here, Kornai (1991), Carson-Berndsen (1991, 1992) and Kirchhoff (1995b) were designed explicitly with a view to spoken language recognition. The other approaches, Kay (1987), Bird & Ellison (1992) and Wiebe (1992), were designed from the point of view of wanting to provide a computational interpretation for an existing phonological theory. In the next section, the implications for spoken language recognition of developments in finite state phonology are discussed.

3.4 IMPLICATIONS FOR SPOKEN LANGUAGE RECOGNITION

It was seen in this chapter that finite state phonology arose as a result of the search for a restriction on the expressive power of generative phonological rules. Further developments showed that it was possible to represent phonological descriptions in terms of finite state devices which worked with regular expressions. It was shown that not only could phonological description be given a computational interpretation, but that this interpretation had the minimum power required for expressing such descriptions. This has positive consequences for spoken language recognition.

Recent work in the area of multi-linear finite state phonology shows that an autosegmental representation can also be interpreted in terms of a finite state automaton. This means that for speech recognition, it is no longer necessary for a phonological segmentation and classification to take place before phonological parsing can commence. Features can and should be detected autonomously and in parallel. In this way, no information which is relevant for phonological processing is lost as is the case in phoneme-based segmentation and classification. Input to phonological parsing is then a multi-linear representation of overlapping features or events. Each feature or event is an autosegment with temporal annotations. The autosegments are not segmented further in terms of phonological segments and therefore there are no strings of the type suggested in Bird & Ellison (1994). The extraction of phonological segments (if indeed these are required) from the autosegmental structure is part of the phonological parsing task. In this way, coarticulation effects in spoken language can be described and processed. The notion of temporality is discussed further in the next chapter.

As mentioned in connection with both the linear encoding model (Kornai, 1991) and the primary tier automaton (Carson-Berndsen, 1991, 1992) above, an automaton for interpreting autosegmental representations performs a check on well-formedness. The representation is accepted or rejected based on the constraints defined in the description. The fact that there may be more information in the input representation than is required by the phonology is not a problem for the primary tier finite automaton since it only requires that its constraints are fulfilled. Furthermore, it is possible to introduce heuristics which allow for constraint relaxation in certain contexts. In this way, parsing becomes more robust. In the rest of this book, it is demonstrated how this approach to multi-linear finite state phonology can be applied for phonological parsing in a spoken language recognition system.

4

THE EVENT CONCEPT IN TIME MAP PHONOLOGY

Time Map Phonology takes an event-based approach to spoken language processing (cf. Carson-Berndsen 1991, 1992a) which is influenced by recent well-motivated developments in the area of nonlinear phonology, in particular autosegmental and constraint-based phonology (cf. Chapter 2). Autosegmental phonology has been criticised (Bird & Ladd, 1991) for not having enough to say about the phonetics/phonology interface. This criticism has also been applied to constraint-based phonology (Carson-Berndsen & Gibbon, 1992), which, although it incorporates a temporal dimension, makes no reference to the relationship between phonological events and actual speech signal tokens. The approach of this book goes further than previous formal interpretations of autosegmental phonology in that phonological relations are explicitly related to intervals in phonetic patterns or actual speech signals and in addition a solution to the projection problem in spoken language recognition is proposed. The task of this chapter is to define the basic concepts required for the event-based phonotactic description in Chapter 5 and for event-relation parsing in Chapter 7.

4.1 MOTIVATION

The motivations for the application of the event concept in spoken language processing are manifold. The main motivation concerns one of the major problems in the processing of speech, namely the *projection problem* at the phonetics/phonology interface. The projection problem was defined in Chapter 1 with respect to the compositionality and variability of spoken language which can only be catered for to a limited extent by the concatenative models of current speech recognition systems which assume a rigid segmentation into non-overlapping units. Furthermore, the projection problem refers to the notion of the native speaker's predictive skill which allows the projection of a finite set of *actual* structures onto a (possibly infinite) set of *potential* structures.

In order to go some way towards solving the projection problem, a more flexible non-segmental approach to spoken language recognition is required which incorporates the notion of compositionality by employing several sources of information simultaneously. Such an approach avoids a strict classification in terms of demisyllables, phonemes or phones. The acoustic front-end does not have to worry about classification over phonological symbols, but can derive autonomous acoustic events from the signal. Such autonomous events are interpreted as an autosegmental representation of sequential and parallel acoustic-phonetic information which is nearer to the signal than the phoneme sequences of more traditional phonologies. The phonetic and phonological knowledge used for processing is represented as well-formedness constraints on autosegmental representations.

A further motivation for the use of events in spoken language recognition concerns the need for an explicit notion of temporality. It is important to distinguish between different temporal domains for speech recognition. Gibbon (1992:93) defines time type levels which co-exist as different perspectives on spoken language utterances. Three of these time types, *category time*, structural or *relative time* and utterance or *absolute time*, are relevant for the discussion which follows. Category time (T_{cat}) refers to categories in an atemporal domain; they have no duration and therefore no direct relation to time. Relational time (T_{rel}) is an abstract temporal domain in which categories are assumed to have duration and can be viewed as intervals with temporal relations between them. Absolute time (T_{abs}) refers to utterance or real signal time. In this domain, categories have a temporal annotation, a direct reference to signal time. In Chapter 2, the difference between traditional distinctive features and event properties with related intervals was described. It is clear that distinctive features are not suitable for phonological parsing as they are represented in category time. However, phonological properties with duration provide the necessary temporal dimension to a categorial linguistic description. They belong to relative time.

In order to be taken seriously in spoken language recognition, phonological description must have a formal interpretation. The lack of formality in phonological theory has long been an issue and it was criticised explicitly by Batóg (1967) who developed a formal axiomatic system using logic and elements of mereology to represent the the informal ideas of Harris (1961). However, this was a segmental approach to phonology and therefore the formal system incorporated only the axioms necessary to describe segmental phenomena. Nonlinear phonology is not segmental and therefore requires a formal interpretation interpretation which goes further than Batóg (1967). Bird & Klein (1990) have shown how autosegmental and articulatory gestural representations can be formally interpreted using events and the axioms of temporal logic. This interpretation allows the explicit introduction of a temporal dimension into the phonological description although constraint-based phonology, as described in Chapter 2, refers to T_{rel}, an abstract temporal domain in terms of intervals and rejects the notion of points. It is shown in the next section that absolute time annotations in terms of endpoints must also be provided in order to relate particular acoustic events to a speech signal token. The emphasis in this book is not the formal interpretation of autosegmental phonology but rather the use of this formalisation in a concrete spoken language recognition system.

4.2 TEMPORAL LOGICS

In the context of spoken language recognition, absolute time must be taken into consideration in connection with a phonological model. In order to abstract away from the explicit temporal annotations which are relevant for token utterances to a phonological level which represents this temporal knowledge more implicitly, a logic which caters for temporal relations is required. This section is concerned with temporal logics which have provided the basis for the work described in the next sections of this chapter.

Allen (1983) addresses the problem of representing temporal knowledge and of temporal reasoning with this knowledge from the perspective of artificial intelligence. He proposes an interval-based temporal logic and a computationally effective reasoning algorithm based on constraint propagation (Allen (1983:832)). Allen discusses the advantages of the interval-based approach as opposed to an approach based on end-points. In particular, he claims that the representation based on time-points is "too uniform and does not facilitate structuring knowledge in a way which is convenient for typical temporal reasoning" (Allen (1983:834)). Allen defines 13 relations based on temporal intervals which are illustrated in Table 4.1.

Table 4.1: 13 temporal relations of Allen (1983)

X before Y	Y after X
X equal Y	
X meets Y	Y is met by X
X overlaps Y	Y is overlapped by X
X during Y	Y contains X
X starts Y	Y is started by X
X finishes Y	Y is finished by X

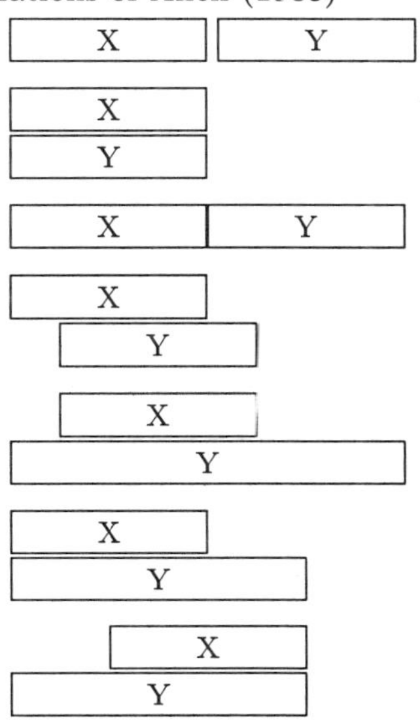

These relations are mutually exclusive. In particular it is important to note that the *overlap* relationship is distinct from the *during* relationship. This distinction will not be made in Section 4.3; temporal overlap will be regarded as a special case of overlap for the purposes of phonological parsing. Allen proposes the use of *reference intervals* in order to minimise the amount of computation required when a new fact is added to the knowledge domain. Reference intervals "group together clusters of intervals for which the temporal constraints between each pair of intervals in the cluster is fully computed." (Allen: 1983:838)). The reference intervals define an indirect relationship to the other intervals in the system. Allen's arguments in favour of an interval-based approach to temporal reasoning are convincing. For this reason, a mapping from the absolute-time domain to the relative (interval-based) time domain is proposed later in the chapter.

Table 4.2: 11 semi-interval relationships of Freksa (1992)

X is older than Y	Y is younger than X	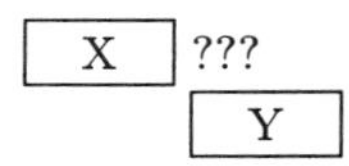
Y is head to head with Y	Y is head to head with X	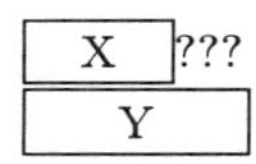
X survives Y	Y is survived by Y	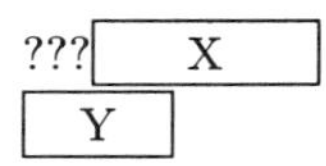
X is tail to tail with Y	Y is tail to tail with X	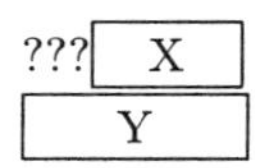
X precedes Y	Y succeeds X	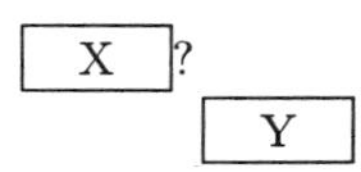
X is a contemporary of Y	Y is a contemporary of X	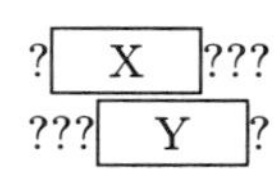
X is born before the death of Y	Y died after birth of X	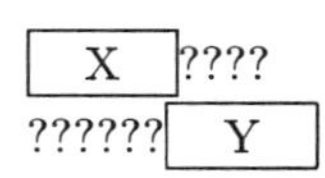

Allen & Hayes (1987) introduce a new relation *meet* into the interval-based approach. They show this single relation to form the basis of the interval-based theory. All relations proposed by Allen (1983) are subsumed by this *meet* relation. They extend the interval-based theory by formally defining the beginnings and endings of intervals which they term *nests*. They distinguish between nests and *moments* which refer to very simple intervals with no internal structure and the role they play with respect to various possible models.

Freksa (1992) addresses the problem of what should be done when only partial information about a domain is available. It may be the case that a situation arises in which uncertainty exists as to which of Allen's 13 mutually exclusive abstract relations hold. Freksa offers a solution based on the notion of *semi-intervals* with a neighbourhood-based inference strategy which he claims is a generalisation of Allen's approach. Semi-intervals denote the beginning and endings of events (equivalent to Allen & Hayes' nests (Freksa (1992:205)). Freksa suggests representing knowledge about time in terms of relationships between semi-intervals. For this 11 semi-interval relationships are required which are presented in Table 4.2.

As will be seen below, what is required for incremental spoken language recognition is an approach which allows for uncertain information about the beginnings and endings of events. The proposal presented in the next section is based on the notion of *open events* since at some particular point in the processing of the token speech signal, the ending may not be known.

4.3 EVENTS AND EVENT RELATIONS

In this section, definitions which are necessary for the understanding of the following chapters are presented. Events will be defined with respect to two domains: the temporal domain (T) and the knowledge domain (K). A temporal domain refers to different perspectives on spoken language utterances. As described in the previous section, *Time Map Phonology* assumes three temporal domains:

- T_{cat}: the category time domain
- T_{rel}: the relative or structural time domain
- T_{abs}: the absolute or utterance time domain

A knowledge domain refers to any linguistic level. In the context of *Time Map Phonology*, there are three knowledge domains:

- $K_{acoustic}$: the acoustic knowledge domain
- $K_{phonetic}$: the phonetic knowledge domain
- $K_{phonological}$: the phonological knowledge domain

The definition space for *Time Map Phonology* is illustrated in Figure 4.1. This classification allows for the provision of phonological explanations in terms of temporality and knowledge. Other approaches provide explanations primarily with respect to one or other of these domains: feature theory, for example, provides knowledge explanations for phonological phenomena whereas metrical phonology provides temporal explanations for phonological phenomena.

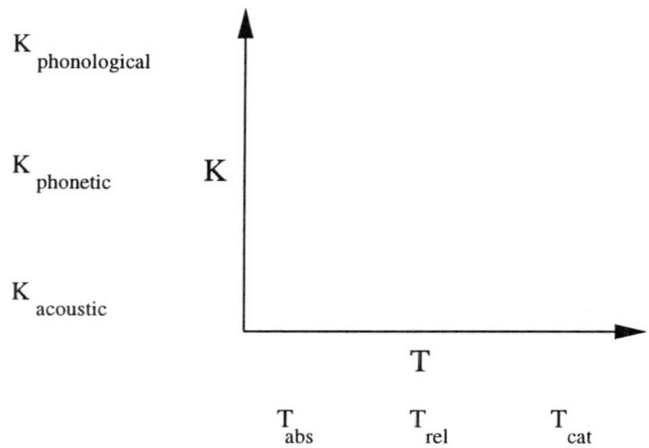

Figure 4.1: Definition space for events in Time Map Phonology

Before considering the temporal and knowledge domains further, a general event definition is presented; this notation will be used throughout this book. The definition corresponds to that given in Bird & Klein (1990:33) and van Benthem (1983:113) except that the terms are different; their terms are given in parentheses.

Definition 4.1 *An event is defined as an ordered pair E = <TR, KR> denoting a temporal representation TR (an interval) and a knowledge representation KR (an event property or description).*

This definition makes no reference to any particular time domain or to any particular knowledge domain. The Bird & Klein (1990) notion of an 'event' may be assigned to the relative time domain (T_{rel}); the knowledge domain is phonology ($K_{phonological}$). This can be seen in the fact that the temporal relations between events are their main concern and that the attributes they work with are phonological in nature. In the context of this book, *TR* can refer to intervals or endpoint annotations. *KR* is not restricted to any particular type of knowledge representation. It can be interpreted as attribute-value structures, unary features, binary features or λ-expressions. Attribute-value structures are used in the examples below.

4.3.1 The Temporal Domain

In this section, emphasis is placed first on the temporal domain; events are defined according to temporal domains where the knowledge domain is specified initially as an abstract parameter K. Events are defined in Section 4.3.2 with respect to the knowledge domains; in this case the temporal domain is specified as an abstract parameter T. In Section 4.4 concrete examples in both domains are provided.

With respect to temporal domains, a distinction is made between *relative-time events* in T_{rel} and *absolute-time events* in T_{abs}. Since it refers to T_{rel}, the relative-time event definition corresponds to the general definition of an event presented above.

Definition 4.2 *A relative-time event is defined as an ordered pair* $\rho\tau\epsilon = < TR, KR >$ *where TR refers to some interval in the time domain* T_{rel} *and KR refers to the knowledge representation in that interval in the domain K.*

A relative-time event can still have a length (Δt) which is part of the TR. This is interpreted as relative duration. This point is not discussed further here (but cf. Chapter 10).

Using axioms of temporal logic, Bird & Klein (1990) define certain relations hold between events. It is these relations which are the basis for a phonological interpretation of the events. The temporal relations hold between relative-time events and are restricted to the time domain T_{rel}; they are regarded as an abstraction from temporal annotations. The temporal relations are overlap, represented by the symbol $^\circ$, precedence, represented by the symbol $\prec$, temporal inclusion, represented by the symbol $\sqsubseteq_t$, and immediate precedence, represented by the symbol $\prec\ ^\circ$. Temporal relations are defined with respect to the following properties: reflexivity, symmetricality and transitivity. A relation R is said to be reflexive if *aRa*. A relation R is said to be symmetric if *aRb* $\Leftrightarrow$ *bRa* and a relation R is said to be transitive if given *aRb* and *bRc*, it is also true that *aRc*.

The set of axioms *RA* governing temporal relations of a set of relative-time events $\rho\tau\epsilon$ is based on those presented in Bird & Klein (1990:39); they have been extended to cater for temporal inclusion and immediate precedence explicitly. The rule of inference is *Modus Ponens*. For any relative-time events $\rho\tau\epsilon_n$

- R1: $\rho\tau\epsilon_i \ ^\circ\ \rho\tau\epsilon_i$
- R2: $\rho\tau\epsilon_i \ ^\circ\ \rho\tau\epsilon_j \rightarrow \rho\tau\epsilon_j \ ^\circ\ \rho\tau\epsilon_i$
- R3: $\rho\tau\epsilon_i \prec \rho\tau\epsilon_j \rightarrow \neg\rho\tau\epsilon_j \prec \rho\tau\epsilon_i$
- R4: $\rho\tau\epsilon_i \prec \rho\tau\epsilon_j \rightarrow \neg\rho\tau\epsilon_i \ ^\circ\ \rho\tau\epsilon_j$
- R5: $\rho\tau\epsilon_i \prec \rho\tau\epsilon_j \wedge \rho\tau\epsilon_j \ ^\circ\ \rho\tau\epsilon_k \wedge \rho\tau\epsilon_k \prec \rho\tau\epsilon_l \rightarrow \rho\tau\epsilon_i \prec \rho\tau\epsilon_l$

- R6: $\rho\tau\epsilon_{\mathrm{i}} \sqsubseteq_{\mathrm{t}} \rho\tau\epsilon_{\mathrm{j}} \rightarrow \rho\tau\epsilon_{\mathrm{i}} \,^{\circ}\, \rho\tau\epsilon_{\mathrm{j}}$
- R7: $\rho\tau\epsilon_{\mathrm{i}} \prec \rho\tau\epsilon_{\mathrm{j}} \wedge \nexists \rho\tau\epsilon_{\mathrm{k}}[\rho\tau\epsilon_{\mathrm{i}} \prec \rho\tau\epsilon_{\mathrm{k}} \wedge \rho\tau\epsilon_{\mathrm{k}} \prec \rho\tau\epsilon_{\mathrm{j}}] \rightarrow \rho\tau\epsilon_{\mathrm{i}} \prec^{\circ} \rho\tau\epsilon_{\mathrm{j}}$

These axioms together with Modus Ponens inference, result in the following conclusions. Overlap is a reflexive, symmetric, and intransitive relation. Precedence is an irreflexive, asymmetric and transitive relation. Temporal inclusion is a reflexive, asymmetric and transitive relation and immediate precedence is an irreflexive, asymmetric and intransitive relation. Note, however, that temporal inclusion does not refer to true simultaneity. It may be the case that A $\sqsubseteq_{\mathrm{t}}$ B and B $\sqsubseteq_{\mathrm{t}}$ A but this is not a property of the relation.

It must also be possible for spoken language processing to relate an event to the absolute time domain ($\mathrm{T}_{\mathrm{abs}}$) in terms of signal time. In order to do this the definition of an event must make some explicit reference to interval endpoints. The following definition of an absolute-time event in $\mathrm{T}_{\mathrm{abs}}$ is given here:

Definition 4.3 *An absolute-time event is defined as an ordered pair* $\alpha\tau\epsilon = << t_{\mathrm{s}}, t_{\mathrm{f}} >, KR >$ *where* t_{s} *is the starting time of the interval and* t_{f} *is the finishing time of the interval in the time domain* T_{abs}. *KR is the knowledge representation of the event in some K of the interval bounded by* t_{s} *and* t_{f}.

These events are, for obvious reasons, also known as temporally-annotated events. Two special cases of absolute-time events are differentiated:

Definition 4.4 *A closed event is an absolute-time event which has both starting and finishing time specifications.*

Definition 4.5 *A open event is an absolute-time event which has no finishing time specification. It is represented as an ordered pair* $<< t_{\mathrm{s}}, >, KR >$.

The closed event is the standard case. An open event refers to the status of an event at a particular time, that is to say, when its finishing time specification is, at some point in time, unknown.

The set of axioms *AA* governing absolute-time events are based on the temporal annotations. They distinguish between closed and open events. The diagrams illustrate the relations between the absolute-time events, similar to the diagrams provided in Bird & Klein (1990:50) and van Benthem (1983: 70). For two closed absolute-time events $\alpha\tau\epsilon_1 = << t_{\mathrm{s1}}, t_{\mathrm{f1}} >, \kappa_1 >$ and $\alpha\tau\epsilon_2 = << t_{\mathrm{s2}}, t_{\mathrm{f2}} >, \kappa_2 >$, the following axioms hold :

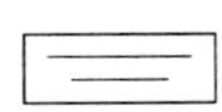

- A1: $\mathrm{t}_{\mathrm{s1}} \leq t_{\mathrm{s2}} \wedge t_{\mathrm{f1}} \geq t_{\mathrm{f2}} \rightarrow \alpha\tau\epsilon_1 \sqsubseteq_{\mathrm{t}} \alpha\tau\epsilon_2$

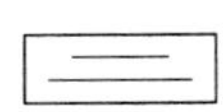

- A2: $\mathrm{t}_{\mathrm{s1}} \geq t_{\mathrm{s2}} \wedge t_{\mathrm{f1}} \leq t_{\mathrm{f2}} \rightarrow \alpha\tau\epsilon_2 \sqsubseteq_{\mathrm{t}} \alpha\tau\epsilon_1$

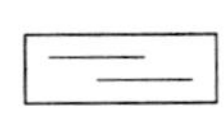

- A3: $\mathrm{t}_{\mathrm{s1}} \leq t_{\mathrm{s2}} \wedge t_{\mathrm{f1}} \geq t_{\mathrm{s2}} \rightarrow \alpha\tau\epsilon_1 \,^{\circ}\, \alpha\tau\epsilon_2$

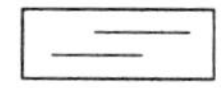

- A4: $\mathrm{t_{s1}} \geq t_{s2} \wedge t_{f1} \leq t_{s2} \rightarrow \alpha\tau\epsilon_1 \circ \alpha\tau\epsilon_2$

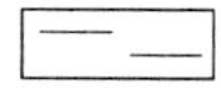

- A5: $\mathrm{t_{f1}} < t_{s2} \rightarrow \alpha\tau\epsilon_1 \prec \alpha\tau\epsilon_2$

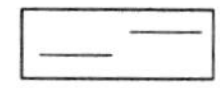

- A6: $\mathrm{t_{f2}} < t_{s1} \rightarrow \alpha\tau\epsilon_2 \prec \alpha\tau\epsilon_1$

Note that further overlap relations can be inferred from the axioms which govern temporal inclusion. These axioms are relevant for closed absolute-time events. It is also necessary to define axioms for *open events*. For two absolute-time events $\alpha\tau\epsilon_1 = << t_{s1}, >, \kappa_1 >$ and $\alpha\tau\epsilon_2 = << t_{s2}, t_{f2} >, \kappa_2 >$, where $\alpha\tau\epsilon_1$ is an open event the following axioms hold:

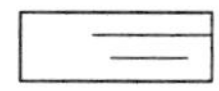

- A7: $\mathrm{t_{s1}} \leq t_{s2} \rightarrow \alpha\tau\epsilon_1 \sqsubseteq_{\mathrm{t}} \alpha\tau\epsilon_2$

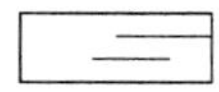

- A8: $\mathrm{t_{s2}} \leq t_{s1} \wedge t_{f2} \geq t_{s1} \rightarrow \alpha\tau\epsilon_1 \circ \alpha\tau\epsilon_2$

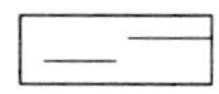

- A9: $\mathrm{t_{s1}} > t_{f2} \rightarrow \alpha\tau\epsilon_2 \prec \alpha\tau\epsilon_1$

Analogously, for two absolute-time events $\alpha\tau\epsilon_1 = << t_{s1}, t_{f1} >, \kappa_1 >$ and $\alpha\tau\epsilon_2 = << t_{s2}, >, \kappa_2 >$, where $\alpha\tau\epsilon_2$ is an open event, the following axioms hold (these axioms are clearly redundant since they correspond to A7 to A9 when $\alpha\tau\epsilon_1$ and $\alpha\tau\epsilon_2$ are reversed; they are provided here for clarity):

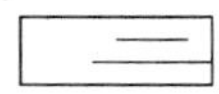

- A10: $\mathrm{t_{s2}} \leq t_{s1} \rightarrow \alpha\tau\epsilon_2 \sqsubseteq_{\mathrm{t}} \alpha\tau\epsilon_1$

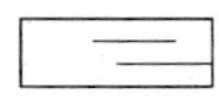

- A11: $\mathrm{t_{s1}} \leq t_{s2} \wedge t_{f1} \geq t_{s2} \rightarrow \alpha\tau\epsilon_1 \circ \alpha\tau\epsilon_2$

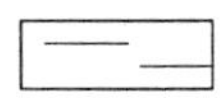

- A12: $\mathrm{t_{s2}} > t_{f1} \rightarrow \alpha\tau\epsilon_1 \prec \alpha\tau\epsilon_2$

Finally, for two open absolute time events $\alpha\tau\epsilon_1 = << t_{s1}, >, \kappa_1 >$ and $\alpha\tau\epsilon_2 = << t_{s2}, >, \kappa_2$, the following two axioms hold:

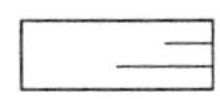

- A13: $\mathrm{t_{s1}} \leq t_{s2} \rightarrow \alpha\tau\epsilon_1 \circ \alpha\tau\epsilon_2$

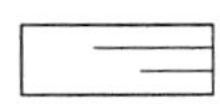

- A14: $\mathrm{t_{s2}} \leq t_{s1} \rightarrow \alpha\tau\epsilon_1 \circ \alpha\tau\epsilon_2$

For obvious reasons, temporal-inclusion and precedence relations cannot be constructed between two open events.

There is a close relationship between absolute-time events and relative time events; the latter is an abstraction of the former, that is to say, it represents the same facts but in a different temporal domain. It is important to note, that this abstraction does not mean that the temporal annotations are no longer present in relative-time events, but that the interval, as a whole, participates in a relationship. It would be possible in the axioms A1 to A14 above to assume that the events on the right hand side of the implication arrow are relative-time events; a temporal relationship exists between the absolute-time events which no longer requires temporal annotations. In the rest of this chapter, where it is not necessary to distinguish between absolute-time events and relative-time events, the more general term *event* is used. The following general representation is used for event relations in the context of this work

Definition 4.6 $\Re = \{\ ^\circ, \prec, \sqsubseteq_t, \prec\ ^\circ\}$ *is the set of possible binary temporal relations between events. The relation between two events* e_i *and* e_j *is represented as ordered triple* $< e_i, \Re', e_j >$ *where* $\Re' \in \{\ ^\circ, \prec, \sqsubseteq_t, \prec\ ^\circ\}$.

This section treated events with respect to the temporal domain. Section 4.3.2 introduces the event concept defined with respect to a knowledge domain.

4.3.2 The Knowledge Domain

In this section, events are defined with respect to the knowledge domain. The knowledge domains which are relevant here are $K_{acoustic}$, $K_{phonetic}$ and $K_{phonological}$. Returning to the general definition of an event as an ordered pair <TR, KR>, the interpretation of KR may be in terms of attribute-value structures, unary or binary features, λ-expressions etc. At the phonetic level, KR refers to unary phonetic properties (e.g. voice) whereas at the phonological level KR is presented in terms of attribute-value structures (multi-valued features) representing features or events on autonomous tiers. As was seen in the discussion of autosegmental, constraint-based and *Time Map Phonology* in Chapter 2, it is necessary to have some way of distinguishing tiers. Bird & Klein (1990) use an index to mark the tier to which a particular attribute refers. The tiers are regarded in the context of this work as generic event attributes defined according to some knowledge domain. A knowledge representation at the phonological level is assumed then to have the structure <TIER: property>. An example of a complete specification for a relative time event in $K_{phonological}$ is $< tr, <$ PHONATION : *voiced* $>>$; analogously an absolute-time event in $K_{phonological}$ is $<< 544, 670 >, <$ PHONATION : *voiced* $>>$.

A basic distinction is drawn between simplex and complex events in a knowledge domain K.

Definition 4.7 *A simplex event is an event in the knowledge domain K, the knowledge representation of which has no internal structure with respect to K.*

Definition 4.8 *A complex event is an event in the knowledge domain K, the knowledge representation of which has been constructed compositionally from simplex events in K using the temporal relations* $\Re$.

Each knowledge domain K has both simplex and complex events. A complex event in one knowledge domain may be a simplex event in a 'higher' knowledge domain, assuming, for *Time Map Phonology* in the spoken language recognition context, that the knowledge domains are ordered as in Figure 4.2.

An event definition in the acoustic domain $K_{acoustic}$ is given in Lancé et al. (1991:6) as the occurrence of a particular structure in the input signal or in a parametric derivation of this signal. In $K_{acoustic}$, acoustic events are usually a function of several parameters and are by no means 'simplex' in this domain. In $K_{phonetic}$, however, an acoustic event is said to be simplex

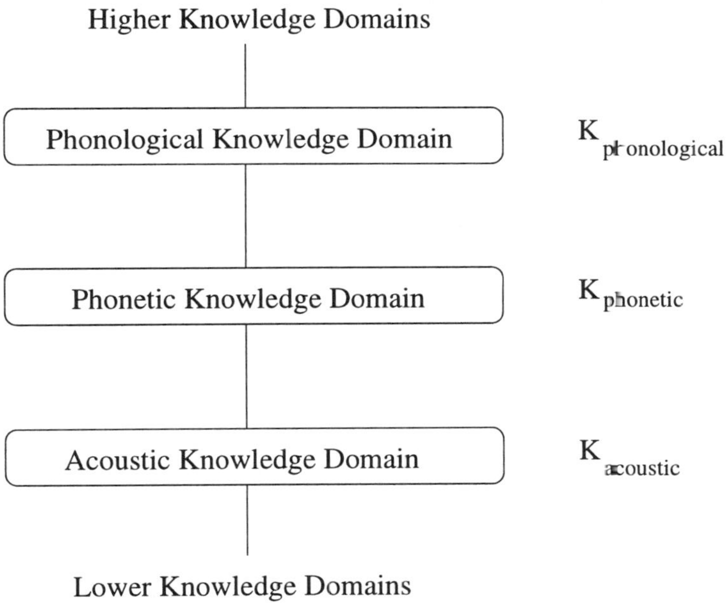

Figure 4.2: Knowledge domains for *Time Map Phonology*

as the internal structure is not transparent. A complex acoustic event in K_{acoustic} is therefore regarded as a simplex (phonetic) event in K_{phonetic}. Similarly, a complex phonetic event in K_{phonetic} is a simplex phonological event in $K_{\text{phonological}}$. Examples of simplex phonetic events are $< TR, <$ *occlusion* $>>$ or $<$ TR, $<$ *syllable−peak* $>>$. Examples of simplex phonological events are $< TR, <$ MANNER : *plosive* $>>$ or $<$ TR, $<$ V-PLACE : *back2front* $>>$. The event property *back2front* refers to a complex feature representing a transition in vowel place qualities in diphthongs (i.e. from a back vowel to a front vowel as in [aɪ]).

A complex event has a transparent structure with respect to a particular knowledge domain. An example of a complex phonological event is $<$ TR, $<$ CODA : *syllable−coda* $>>$ (cf. discussion of the syllable in Chapter 5) which is constructed compositionally on the basis of the overlap and precedence of simplex phonological events in domain $K_{\text{phonological}}$. An example of a complex phonetic event is $<$ TR, $<$ V-PLACE : *back2front* $>>$ or $<$ TR, $<$ MANNER : *plosive* $>>$. These events are based on overlap and precedence of simplex phonetic events in the domain K_{phonetic}. $<$ TR, $<$ MANNER : *plosive* $>>$, for instance, is constructed compositionally on the basis of the simplex $<$ TR, $<$ *occlusion* $>>$, $<$ TR, $<$ *release* $>>$, and $<$ TR, $<$ *noise* $>>$ events. The set of phonetic and phonological events used in this book are provided in the next section.

Before moving to a more detailed discussion of the phonetic and phonological events relevant for German, it is necessary to define the notion of an event lattice.

Definition 4.9 *An event lattice is defined as an ordered triple* $\mathcal{EL} = <\mathcal{E}, {}^{\circ}, \prec>$ *where* $\mathcal{E}$ *is a set of events in some knowledge domain K,* $^{\circ}$ *is the set of overlap relations existing between elements of* $\mathcal{E}$ *and* $\prec$ *is the set of precedence relations existing between elements of* $\mathcal{E}$*.*

This definition is similar to that of a phonological event structure provided in Bird & Klein (1990:43). However, it differs in that an event lattice refers to *all* possible overlap and precedence relations between events of $\mathcal{E}$. An event lattice refers to a token rather than a type description. Theoretically, three interpretations of $\mathcal{EL}$ are possible with respect to overlap information. Overlapping information can be regarded as disjunctive or conjunctive allowing for the following three distinctions in *Time Map Phonology*:

- Disjunctive event lattices (corresponding to the traditional word or phoneme lattices of speech language recognition systems) where overlapping information refers to alternative events.
- Conjunctive event lattices where overlapping information refers to cooccurring events.
- Competitive conjunctive event lattices where overlapping information may be either disjunctive (competing alternative events) or conjunctive (cooccurring events).

This section has provided general definitions of events which are relevant for event-based *tactic* (syntagmatic) descriptions. Events were defined with respect to temporal and knowledge domains. It was also shown how events from different temporal domains may be related via axioms governing temporal relations. It is shown in Chapter 5 how event descriptions (*types*) for both the phonetic and phonological domains can be defined in $\mathrm{T}_{\mathrm{rel}}$ and how events from the different knowledge domains can be related declaratively. In Chapter 7, it is shown how the event descriptions (*types*) are utilised to constrain the interpretation of events in *token* utterances. First, however, it is necessary to define phonological and phonetic events in the respective knowledge domains. The next section is concerned with events in the knowledge domains $\mathrm{K}_{\mathrm{phonetic}}$ and $\mathrm{K}_{\mathrm{phonological}}$ for German.

4.4 PHONETIC AND PHONOLOGICAL EVENTS FOR GERMAN

This section is concerned with events in the knowledge domains of *Time Map Phonology*. This book is concerned primarily with $\mathrm{K}_{\mathrm{phonetic}}$ and $\mathrm{K}_{\mathrm{phonological}}$. Although the acoustic event catalogue is presented here as simplex phonetic events, nothing will be said about how such acoustic events are derived from the signal. Research on detection of autonomous acoustic events for German in $\mathrm{K}_{\mathrm{acoustic}}$, which serve as primitives for *Time Map Phonology* is described in Lancé et al. (1991), Faust & Möbius (1992), Hauenstein & Hübener (1992), Hübener (1993) and Hübener & Carson-Berndsen (1994). In this section, a set of phonetic and phonological events which are relevant for the description of German are defined. In both domains, knowledge representations in terms of attribute-value structures are assumed. As was seen in the examples of the previous section, simplex and complex events are not restricted to conventional features, but rather can refer to any categories which are considered relevant in the context of spoken language recognition.

4.4.1 The Phonetic Domain: $\mathrm{K}_{\mathrm{phonetic}}$

The discussion begins with the definition of the event set in $\mathrm{K}_{\mathrm{phonetic}}$ which is relevant for the description of German. The phonetic event set $\mathcal{E}_{\mathrm{phonetic}}$is based on the acoustic events defined in Faust & Möbius (1992) and is restricted initially to those acoustically detectable events which are relevant for German. The set of properties has been extended as described in Hübener & Carson-Berndsen (1994); they are given the in Table 4.3 together with the abbreviations used in the screenshots in Chapter 7 (the abbreviations are based on the German terms). The voice, noise and pause properties correspond to the source features of Fant (1973), the other properties apply to vowels and consonants respectively (cf. Faust & Möbius, 1992 and Carson-Berndsen, 1991).

On the basis of the properties in Table 4.3. the phonetic event set $\mathcal{E}_{\mathrm{phonetic}}$ is defined. These events should be regarded as types rather than tokens. In each case, the temporal representations refer to intervals in $\mathrm{T}_{\mathrm{rel}}$ and are thus marked by $\mathrm{I}_{\mathrm{rel}}$. $\mathcal{E}_{\mathrm{phonetic}}$ has elements of the form $< I_{\mathrm{rel}}, < property >>$ for each of the attribute-value structures defined in Table 4.3. $\mathcal{E}_{\mathrm{phonetic}}$ is assumed to be the set of simplex phonetic events. Complex phonetic events in $\mathrm{K}_{\mathrm{phonetic}}$ correspond to the set of simplex phonological events in $\mathrm{K}_{\mathrm{phonological}}$ which are defined in Section 4.4.2.

$$\begin{aligned}\mathcal{E}_{\mathrm{phonetic}} \quad = \{ \quad & < I_{\mathrm{rel}}, < voice >>, \\ & < I_{\mathrm{rel}}, < noise >>, \\ & < I_{\mathrm{rel}}, < pause >>, \\ & < I_{\mathrm{rel}}, < syllable - peak >>, \\ & < I_{\mathrm{rel}}, < vowellike >>, ...\}\end{aligned}$$

Table 4.3: Attribute-value table for K_{phonetic}

Phonetic Features	*Abbreviation*
pause	pa
vowel-transition	tv
dark-vowel	vd
round-vowel	vr
low-vowel	vt
[a]-vowel	va
central-vowel	vz
nonround-vowel	vu
light-vowel	vh
high-vowel	vg
mid-vowel	vm
vowellike	vo
syllable-peak	sk
nasality	na
lateral	la
voice	sh
noise	gh
occlusion	op
release	ol
frication	fr
glottal	gl
uvular	uv
velar	ve
palatal	pl
palato-alveolar	po
apical	ap
labial	lb

On the basis of $\mathcal{E}_{\text{phonetic}}$ it is possible to represent phonetic segments; these remain autonomous from a phonological point of view; they correspond to autosegments on phonological tiers. Phonetic segments are represented in terms of an ordered triple $<C_{\text{phonetic}}, \Re^{\circ}, \Re \prec>$ where $C_{\text{phonetic}} \subseteq \mathcal{E}_{\text{phonetic}}$ is a set of phonetic events, $\Re^{\circ}$ is the set of overlap relations between members of C_{phonetic} and $\Re \prec$ is the set of precedence relations between members of C_{phonetic}. This tuple refers to representation and is not a definition of a phonetic segment. The intervals of events in C_{phonetic} are assumed to be in T_{rel} and the differing intervals are marked by a subscript to the I. Relations have the format as defined in Section 4.3.1. Examples of possible phonetic segments are:

$$
\begin{aligned}
[\text{plosive}] = \\
&< \{< I_i, < \textit{occlusion} >>, \\
&< I_j, < \textit{release} >>, \\
&< I_k, < \textit{noise} >>\}, \\
&\{< I_j, {}^\circ, \mathrm{I_k} >\}, \\
&\{< I_i, \prec, \mathrm{I_j} >\} >
\end{aligned}
$$

$$
\begin{aligned}
[\text{back2front}] = \\
&< \{< I_i, < \textit{dark} - \textit{vowel} >>, \\
&< I_j, < \textit{light} - \textit{vowel} >>\}, \\
&\{\}, \\
&\{< I_i, \prec, \mathrm{I_j} >\} >
\end{aligned}
$$

The autosegment *plosive* occurs on the autonomous tier MANNER in $\mathrm{K_{phonological}}$. The autosegment *back2front* occurs on the autonomous tier V-PLACE in $\mathrm{K_{phonological}}$. Other phonetic segments are defined in Chapter 5.

An example of a phonetic event description of a token of the German utterance *ich bräuchte* [ɪçbʁɔɪçtə] is provided in Figure 4.3. Only the tiers (attributes) relevant for this example are given. The simplex phonetic events in Figure 4.3 are given below as absolute-time events. The temporal annotations were provided by the *Institut für Kommunikationsforschung und Phonetik*, University of Bonn. Each absolute-time event has been given a unique label $\alpha\tau\epsilon_i$.

$$
\begin{aligned}
&\alpha\tau\epsilon_1{:=}<< 546, 625 >, < \textit{voice} >> \\
&\alpha\tau\epsilon_2{:=}<< 545, 548 >, < \textit{syllable} - \textit{peak} >> \\
&\alpha\tau\epsilon_3{:=}<< 548, 552 >, < \textit{occlusion} >> \\
&\alpha\tau\epsilon_4{:=}<< 548, 698 >, < \textit{light} - \textit{vowel} >> \\
&\alpha\tau\epsilon_5{:=}<< 549, 598 >, < \textit{vowellike} >> \\
&\alpha\tau\epsilon_6{:=}<< 550, 554 >, < \textit{release} >> \\
&\alpha\tau\epsilon_7{:=}<< 585, 662 >, < \textit{frication} >> \\
&\alpha\tau\epsilon_8{:=}<< 593, 798 >, < \textit{noise} >> \\
&\alpha\tau\epsilon_9{:=}<< 670, 726 >, < \textit{occlusion} >> \\
&\alpha\tau\epsilon_{10}{:=}<< 725, 871 >, < \textit{voice} >> \\
&\alpha\tau\epsilon_{11}{:=}<< 726, 734 >, < \textit{release} >> \\
&\alpha\tau\epsilon_{12}{:=}<< 728, 798 >, < \textit{frication} >> \\
&\alpha\tau\epsilon_{13}{:=}<< 782, 956 >, < \textit{vowellike} >> \\
&\alpha\tau\epsilon_{14}{:=}<< 798, 892 >, < \textit{dark} - \textit{vowel} >> \\
&\alpha\tau\epsilon_{15}{:=}<< 824, 948 >, < \textit{vowel} - \textit{transition} >> \\
&\alpha\tau\epsilon_{16}{:=}<< 864, 866 >, < \textit{syllable} - \textit{peak} >> \\
&\alpha\tau\epsilon_{17}{:=}<< 907, 950 >, < \textit{light} - \textit{vowel} >> \\
&\alpha\tau\epsilon_{18}{:=}<< 926, 1034 >, < \textit{frication} >>
\end{aligned}
$$

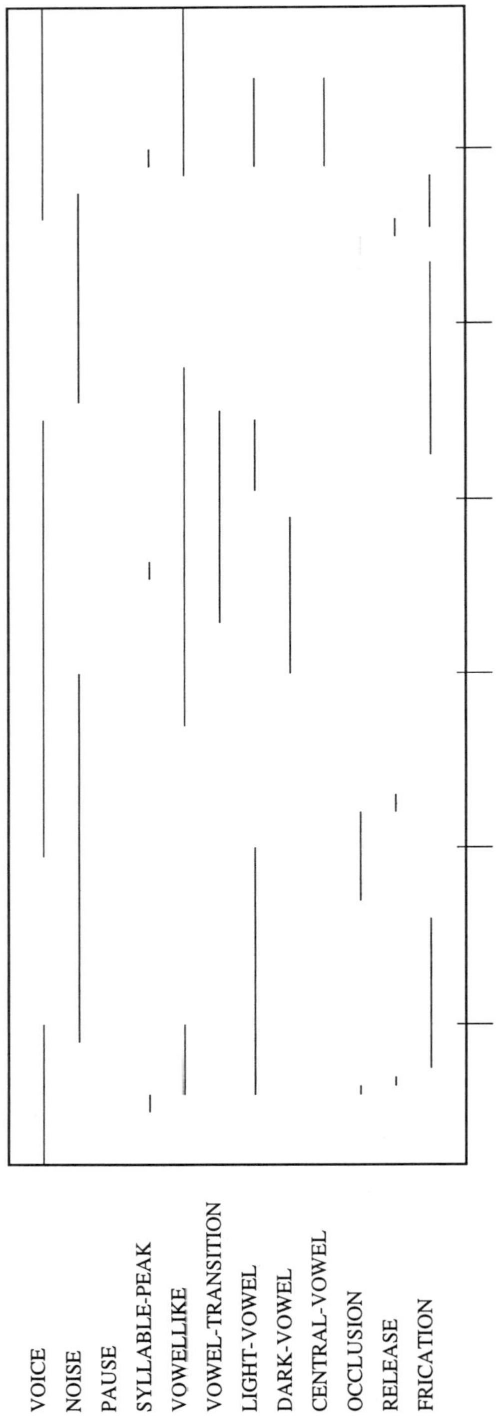

Figure 4.3: Graphical representation of phonetic events for the utterance [ɪçbʁɔɪçtə] *ich bräuchte*

$\alpha\tau\epsilon_{19}$:=<< 939, 1087 >, < *noise* >>
$\alpha\tau\epsilon_{20}$:=<< 1036, 1062 >, < *occlusion* >>
$\alpha\tau\epsilon_{21}$:=<< 1063, 1066 >, < *release* >>
$\alpha\tau\epsilon_{22}$:=<< 1066, 1088 >, < *frication* >>
$\alpha\tau\epsilon_{23}$:=<< 1066, >, < *voice* >>
$\alpha\tau\epsilon_{24}$:=<< 1081, >, < *vowellike* >>
$\alpha\tau\epsilon_{25}$:=<< 1081, 1136 >, < *light − vowel* >>
$\alpha\tau\epsilon_{26}$:=<< 1081, 1127 >, < *central − vowel* >>
$\alpha\tau\epsilon_{27}$:=<< 1094, 1098 >, < *syllable − peak* >>

4.4.2 The Phonological Domain: $K_{\text{phonological}}$

The event set for $K_{\text{phonological}}$ is defined analogously to that for K_{phonetic}. The relevant attribute-value structures (tier/property structures) are defined in Table 4.4. The event properties have been defined with respect to the IPA classification. On the basis of the tiers and properties in Table 4.4. a phonological

Table 4.4: Attribute-value Table for $K_{\text{phonological}}$

Attribute/Tier	*Value/Property*
PHONATION	{voiced, voiceless}
MANNER	{plosive, fricative, nasal, lateral, affricate, vowellike, diphthong}
PLACE	{labial, apical, palato-alveolar, velar, palatal, uvular, glottal}
V-PLACE	{front, back, central, back2front, front2central}
LENGTH	{long, short}
HEIGHT	{high, low, mid}
ROUNDNESS	{round, nonround}
OPENNESS	{open, close}

event set is defined. The events in the phonological event set should be regarded as types rather than tokens. In each case, the temporal representations refer to intervals in T_{rel} and are marked by I_{rel}. The phonological event set $\mathcal{E}_{\text{phonological}}$ has elements of the form $< I_{\text{rel}}, <$ TIER : *property* >> for each of the attribute-value structures defined in Table 4.4. Here again, $\mathcal{E}_{\text{phonological}}$ is assumed to be the set of simplex phonological events. Complex phonological events are syllable and syllable-part events which define the set of simplex events for a higher knowledge domain.

$\mathcal{E}_{\text{phonological}} = \{$ $< I_{\text{rel}}, <$ PHONATION : *voiced* >>,
$< I_{\text{rel}}, <$ PHONATION : *voiceless* >>,
$< I_{\text{rel}}, <$ MANNER : *plosive* >>,

$< I_{\text{rel}},< \text{MANNER} : fricative >>,$
$< I_{\text{rel}}, < \text{MANNER} : nasal >>,$
$< I_{\text{rel}}, < \text{MANNER} : lateral >>,...\}$

Recall that simplex phonological events correspond to complex phonetic events. Some of the events provide explanations of phenomena in terms of knowledge. For example, the simplex phonological event $< I_{\text{rel}}, < \text{V-PLACE} : back2front >>$ describes the change of place within diphthongs in terms of knowledge rather that in temporal terms.

Phonological segments can be represented in terms of overlap relations between the attribute-value structures in Figure 4.3. A selection of phonological segments for German is given here. The segments are represented as ordered pairs $< C_{\text{phonological}}, \Re^{\circ} >$ where $\text{C}_{\text{phonological}} \subseteq \mathcal{E}_{\text{phonological}}$ and $\Re^{\circ}$ is the set of overlap relations between elements of $\text{C}_{\text{phonological}}$. It should be noted that this is a representation and not a definition of a phonological segment. The intervals of events in $\text{C}_{\text{phonological}}$ are assumed to be in T_{rel} and the differing intervals are marked by a subscript to the I. The relations are represented as defined in Section 4.3.1. Note that in these representation $\Re^{\circ}$ is the full set of overlap relations between all the elements of $\text{C}_{\text{phonological}}$ i.e. the overlap relations are fully specified; in such cases, $\Re^{\circ}{}_{\text{i...n}}$ may be substituted.

/ə/ $=< \{< I_{\text{i}}, < \text{PHONATION} : voiced >>, < I_{\text{j}},< \text{MANNER} : vowellike >>, < I_{\text{k}}, < \text{V-PLACE} : central >>, < I_{\text{l}}, < \text{LENGTH} : short >>, < I_{\text{m}},< \text{HEIGHT} : mid >>, < I_{\text{n}}, < \text{ROUNDNESS} : nonround >>\}, \{< I_{\text{i}}, ^{\circ},\text{I}_{\text{j}} >, < I_{\text{j}}, ^{\circ},\text{I}_{\text{k}} >, < I_{\text{i}}, ^{\circ},\text{I}_{\text{k}} >, < I_{\text{i}}, ^{\circ},\text{I}_{\text{l}} >, < I_{\text{i}}, ^{\circ},\text{I}_{\text{m}} >, < I_{\text{i}}, ^{\circ},\text{I}_{\text{n}} >, < I_{\text{j}}, ^{\circ},\text{I}_{\text{l}} >, < I_{\text{j}}, ^{\circ},\text{I}_{\text{m}} >, < I_{\text{j}}, ^{\circ},\text{I}_{\text{n}} >, < I_{\text{k}}, ^{\circ},\text{I}_{\text{l}} >, < I_{\text{k}}, ^{\circ},\text{I}_{\text{m}} >, < I_{\text{k}}, ^{\circ},\text{I}_{\text{n}} >, < I_{\text{l}}, ^{\circ},\text{I}_{\text{m}} >, < I_{\text{m}}, ^{\circ},\text{I}_{\text{n}} >, < I_{\text{l}}, ^{\circ},\text{I}_{\text{n}} >\} >$

/m/ $=< \{< I_{\text{i}}, < \text{PHONATION} : voiced >>, < I_{\text{j}}, < \text{MANNER} : nasal >>, < I_{\text{k}}, < \text{PLACE} : labial >>\}, \{< I_{\text{i}}, ^{\circ},\text{I}_{\text{j}} >, < I_{\text{j}}, ^{\circ},\text{I}_{\text{k}} >, < I_{\text{i}}, ^{\circ},\text{I}_{\text{k}} >\} >$

/p/ $=< \{< I_{\text{i}}, < \text{PHONATION} : voiceless >>, < I_{\text{j}}, < \text{MANNER} : plosive >>, < I_{\text{k}}, < \text{PLACE} : labial >>\}, \{< I_{\text{i}}, ^{\circ},\text{I}_{\text{j}} >, < I_{\text{j}}, ^{\circ},\text{I}_{\text{k}} >, < I_{\text{i}}, ^{\circ},\text{I}_{\text{k}} >\} >$

/ts/ $=< \{< I_{\text{i}}, < \text{PHONATION} : voiceless >>, < I_{\text{j}}, < \text{MANNER} : affricate >>, < I_{\text{k}}, < \text{PLACE} : apical >>\}, \{< I_{\text{i}}, ^{\circ},\text{I}_{\text{j}} >, < I_{\text{j}}, ^{\circ},\text{I}_{\text{k}} >, < I_{\text{i}}, ^{\circ},\text{I}_{\text{k}} >\} >$

An example of a phonological event description of a token of the German utterance *ich bräuchte* /ɪçbʁɔɪçtə/ is given in Figure 4.4. A set of possible absolute-time events for this example are given below. Each absolute-time event has been given a unique label $\alpha\tau\epsilon_{\text{i}}$; only the label subscripts are given in the graphical representation of Figure 4.4. The temporal annotations are an approximation based on the output of a phonological event generation tool (GENEVE, Carson-Berndsen, 1992d) which generates labelled event data based on phonemic representations. It is 'idealised' data since the synchronisation function is the same across tiers; it is presented here purely for illustration purposes. Real data has much more overlap, it is underspecified, and it has competing hypotheses. In figures 4.5 and 7.3 screenshots showing acoustic and phonological events for identical sections an utterance for labelled data and real data respectively. Competing hypotheses are indicated in 4.6 by broader lines which appear in different colours in the computer visualisation. These are represented by varying shades of grey in the figure.

$\alpha\tau\epsilon_{1}{:=}<< 549, 600 >, < \textsc{phonation} : \mathit{voiced} >>$
$\alpha\tau\epsilon_{2}{:=}<< 549, 600 >, < \textsc{manner} : \mathit{vowellike} >>$
$\alpha\tau\epsilon_{3}{:=}<< 549, 600 >, < \textsc{v-place} : \mathit{front} >>$
$\alpha\tau\epsilon_{4}{:=}<< 549, 600 >, < \textsc{length} : \mathit{short} >>$
$\alpha\tau\epsilon_{5}{:=}<< 549, 600 >, < \textsc{height} : \mathit{high} >>$
$\alpha\tau\epsilon_{6}{:=}<< 549, 600 >, < \textsc{roundness} : \mathit{nonround} >>$
$\alpha\tau\epsilon_{7}{:=}<< 549, 600 >, < \textsc{openness} : \mathit{close} >>$
$\alpha\tau\epsilon_{8}{:=}<< 600, 724 >, < \textsc{phonation} : \mathit{voiceless} >>$
$\alpha\tau\epsilon_{9}{:=}<< 600, 724 >, < \textsc{manner} : \mathit{fricative} >>$
$\alpha\tau\epsilon_{10}{:=}<< 600, 724 >, < \textsc{place} : \mathit{palatal} >>$
$\alpha\tau\epsilon_{11}{:=}<< 724, 952 >, < \textsc{phonation} : \mathit{voiced} >>$
$\alpha\tau\epsilon_{12}{:=}<< 724, 746 >, < \textsc{manner} : \mathit{plosive} >>$
$\alpha\tau\epsilon_{13}{:=}<< 724, 746 >, < \textsc{place} : \mathit{labial} >>$
$\alpha\tau\epsilon_{14}{:=}<< 746, 782 >, < \textsc{manner} : \mathit{fricative} >>$
$\alpha\tau\epsilon_{15}{:=}<< 746, 782 >, < \textsc{place} : \mathit{uvular} >>$
$\alpha\tau\epsilon_{16}{:=}<< 782, 952 >, < \textsc{manner} : \mathit{diphthong} >>$
$\alpha\tau\epsilon_{17}{:=}<< 782, 952 >, < \textsc{v-place} : \mathit{back2front} >>$
$\alpha\tau\epsilon_{18}{:=}<< 782, 952 >, < \textsc{roundness} : \mathit{round2nonround} >>$
$\alpha\tau\epsilon_{19}{:=}<< 952, 1081 >, < \textsc{phonation} : \mathit{voiceless} >>$
$\alpha\tau\epsilon_{20}{:=}<< 952, 1062 >, < \textsc{manner} : \mathit{fricative} >>$
$\alpha\tau\epsilon_{21}{:=}<< 952, 1062 >, < \textsc{place} : \mathit{palatal} >>$
$\alpha\tau\epsilon_{22}{:=}<< 1062, 1081 >, < \textsc{manner} : \mathit{plosive} >>$
$\alpha\tau\epsilon_{23}{:=}<< 1062, 1081 >, < \textsc{place} : \mathit{apical} >>$
$\alpha\tau\epsilon_{24}{:=}<< 1081, \ >, < \textsc{phonation} : \mathit{voiced} >>$
$\alpha\tau\epsilon_{25}{:=}<< 1081, \ >, < \textsc{manner} : \mathit{vowellike} >>$
$\alpha\tau\epsilon_{26}{:=}<< 1081, 1155 >, < \textsc{v-place} : \mathit{central} >>$
$\alpha\tau\epsilon_{27}{:=}<< 1081, 1155 >, < \textsc{length} : \mathit{short} >>$
$\alpha\tau\epsilon_{28}{:=}<< 1081, 1155 >, < \textsc{height} : \mathit{mid} >>$
$\alpha\tau\epsilon_{29}{:=}<< 1081, 1155 >, < \textsc{roundness} : \mathit{nonround} >>$
$\alpha\tau\epsilon_{30}{:=}<< 1081, 1155 >, < \textsc{openness} : \mathit{close} >>$

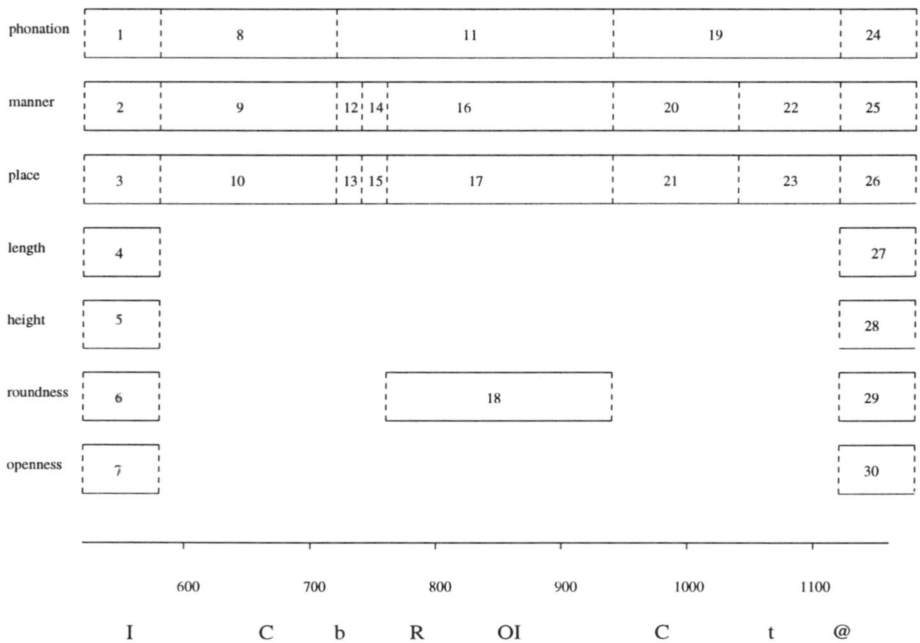

Figure 4.4: A graphical representation of stylised phonological events for the utterance /ɪçbʁɔɪçtə/ *ich bräuchte*

This chapter was concerned with certain declarative aspects of event descriptions. The relationship between phonological events and phonetic/acoustic events is described in the next chapter together with event descriptions (*types*) for the phonetic and phonological description of German. These event descriptions express constraints on possible event structures. The procedural aspects of event processing are discussed in connection with an implementation in Chapter 7.

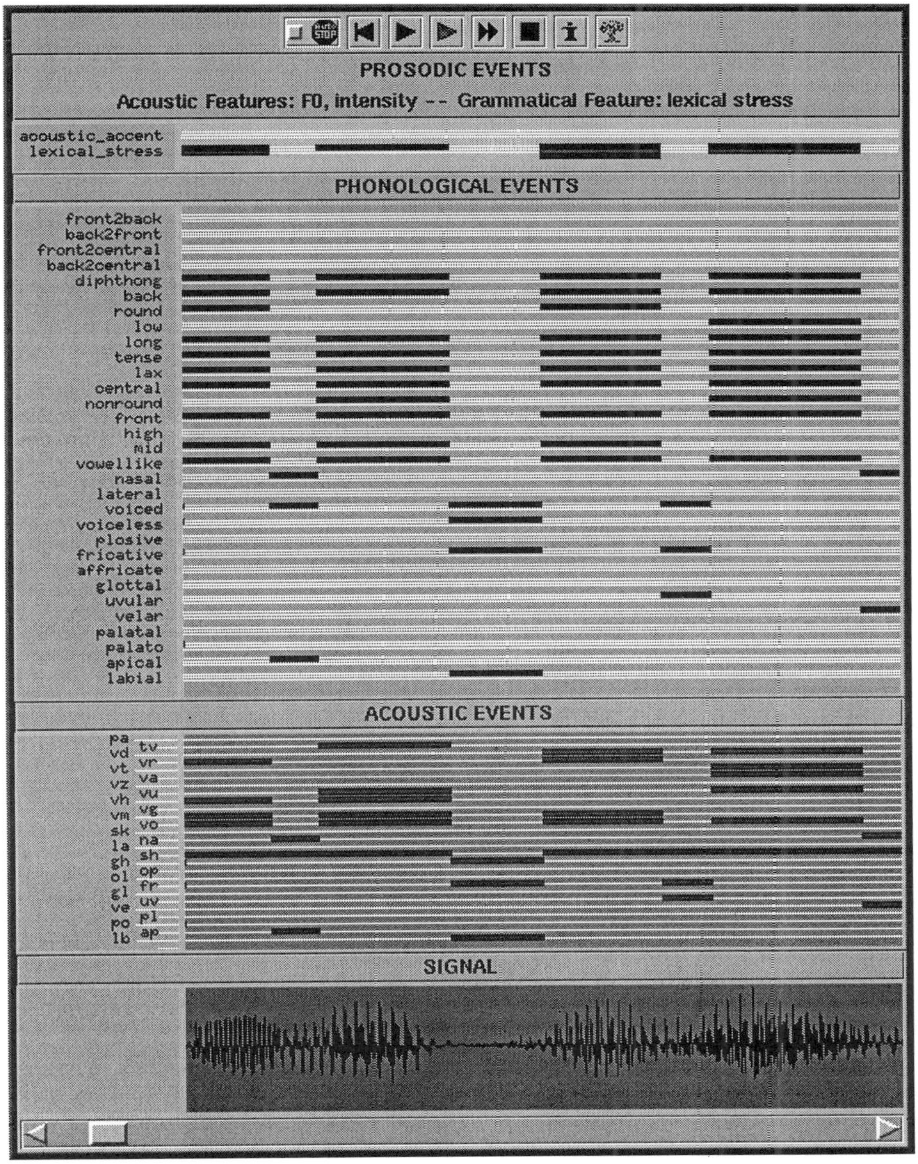

Figure 4.5: A screenshot showing the acoustic and phonological events for labelled data

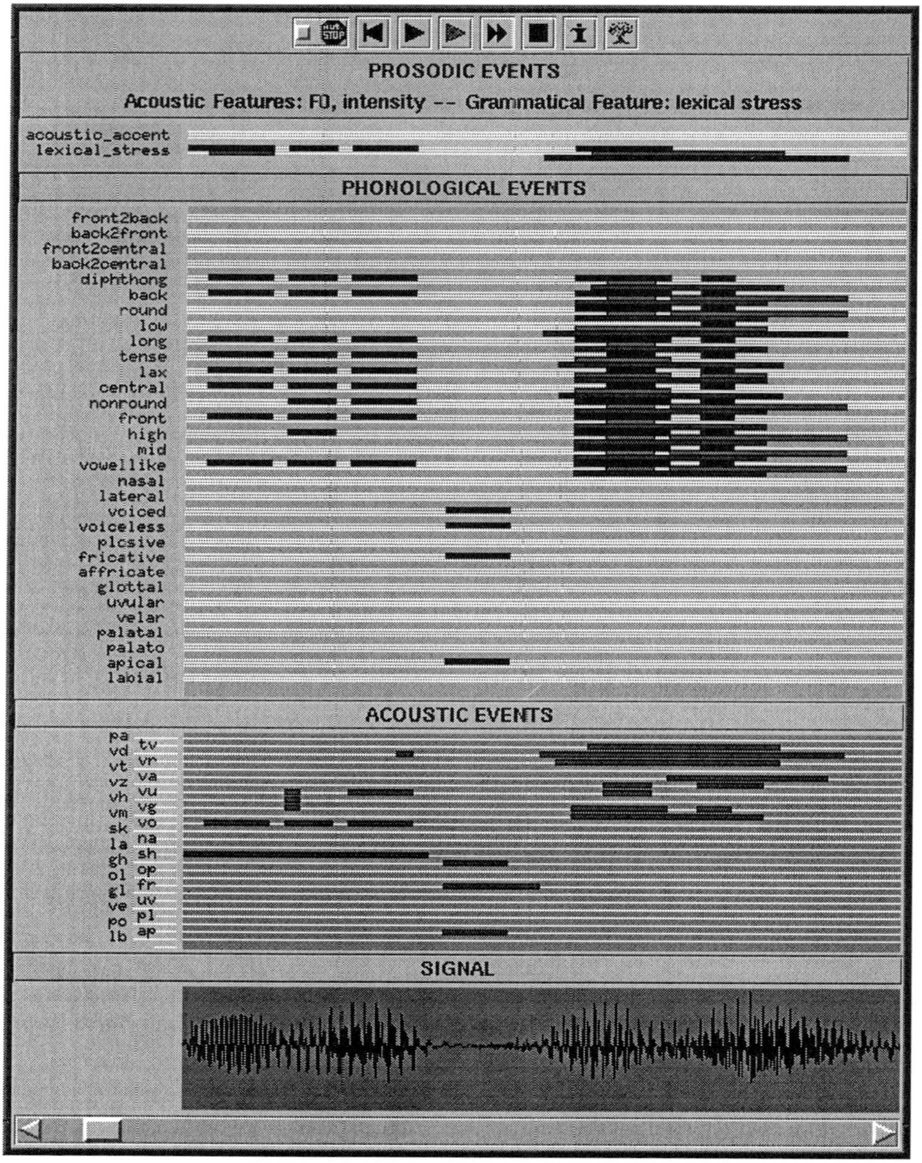

Figure 4.6: A screenshot showing the acoustic and phonological events for real data

5

PHONOTACTIC DESCRIPTIONS AND THEIR REPRESENTATION

Statements of permissible combinations of phonemes in a language are known as its phonotactics. Such statements can be formulated as constraints on well-formedness of phonological structures within the domain of declarative phonology (cf. Chapter 2). However, as will be seen below, although phonemes may be the heuristic starting point for such descriptions, they are not required to act as primitives. In the following sections, adequacy criteria for phonotactic description are introduced and discussed with respect to possible phonotactic representations. A linear segmental network representation formalism for phonotactic description is then formally defined from both a declarative and procedural perspective in Section 5.3. In Section 5.6, this representation formalism is then extended to cater for the multi-linear temporal event descriptions of *Time Map Phonology.* The network representation formalism for event descriptions serves as a basis for the definition of syllable event networks (phonotactic descriptions) and complex phonetic event networks for German in Sections 5.6 and 5.7.

5.1 ADEQUACY CRITERIA FOR PHONOTACTIC DESCRIPTION

Halle (1964) in a revision of a 1958 article discusses requirements of a framework for characterising linguistic descriptions of speech, in particular the requirement that in addition to describing the observed data accurately, the representation provides "reasonable, fruitful, insightful, and simple descriptions of the relevant facts" (Halle, 1964:324). Chomsky (1964:30) introduces levels of adequacy in linguistic description. He discusses the criteria for observational, descriptive and explanatory adequacy in connection with the phonological component of a grammar, presenting as an example the phenomena of accidental (idiosyncratic) and systematic gaps in the lexicon. A phonological grammar must be able to cater for the fact that although there is no attestation of the form /blɪk/ in the English language, a native speaker would still classify this form as a possible word of the language; it is purely accidental that this form does not exist. The fact that the form /ftɪk/ does not exist in the lexicon is systematic, however, since a native speaker would not consider this initial combination to be possible in the language.

In line with the criteria for declarative adequacy of computational phonological models defined in Chapter 2, the following evaluation criteria should be taken into account with respect to phonotactic description:

- *observational adequacy*: in particular the ability to cater for *actual* syllables and phonological words of the language
- *descriptive adequacy*: in particular the ability to generalise beyond *actual* to *potential* syllables and phonological words thereby distinguishing accidental from systematic gaps in the language
- *explanatory adequacy*: in particular the ability to provide a principled reason for selecting the correct description.

Unfortunately, criteria for linguistic adequacy have tended to be neglected in past phonotactic descriptions. Descriptions are either too restricted, that is to say they only describe *actual* forms, or they overgeneralise, describing forms which would be considered ill-formed by a native speaker of a language. In the context of a spoken language recognition system, a linguistic description should provide a notion of well-formedness of structure, it should be able to cater for idiosyncratic gaps in the lexicon, since neologisms and borrowings will tend to conform to the phonotactics of the language. Formally, the linguistic description should be sound and complete (i.e. a model).

5.2 DESCRIBING PHONOTACTIC DATA

There are a number of issues, in addition to the adequacy criteria mentioned, which should be discussed in connection with phonotactic representation. These concern the notions of domain, finiteness and linearity. In the past, the main emphasis in the study of phonotactics concerned not the appropriateness of the representation mechanism employed but rather whether the syllable, the morpheme or the word was the relevant domain for the description. Much of this discussion arose from the fact that different phonologies assumed different domains to the exclusion of others. Generative phonology, for example, uses no syllable boundaries in phonological rules. Well-formedness constraints on morphemes were provided in the lexicon in the form of *morpheme structure rules* (Halle, 1959). The proponents of Natural Generative Phonology (Vennemann, 1972; Hooper, 1976) on the other hand, claim that phonological rules could be stated more simply within the syllable domain. As a result, there has been no unified approach to phonotactic description within phonology as a discipline.

The set of possible words in natural languages, analogous to the set of sentences, is infinite. The set of possible syllables, however, is not; they can be calculated on the basis of how many phonemes can appear in which position of the syllable. An observationally adequate description of phonotactic data, then, could be in terms of a list of all actual or possible syllables. However, this set, although finite, is so large (around 250 000 based on experiments using a syllable network for German) that a list is no longer suitable. Furthermore, a list would not make any generalisation over structures other than the syllable itself. Syllable substructures such as onset (initial consonant clusters), peak (vowels) and codas (final consonant clusters) would have no relevance. However, psycholinguistic experiments show that a native speaker respects the unity of onsets and rhymes, that is to say, a native speaker recognises a syllable substructure rhyme (cf. Treiman, 1986 for evidence on English), which consists of a peak and coda, and that there is a closer relationship between these than between, say, onset and peak. It has also been claimed that there are in fact restrictions which go beyond these substructures. e.g there is no English syllable of the form sCVC where the Cs have the same coronal specification. However, Davis (1992) presents evidence on the basis of a computerised lexicon which refutes these claims. Selkirk (1982) discusses the immediate constituent (IC) principle of phonotactics and proposes an underlying hierarchical organisation of syllable structure into onset and rhyme which further branches into peak and coda (cf. Pike & Pike, 1948 for an earlier proposal of this type of syllable organisation). This underlying hierarchical structure expresses a distributional "chunking" in phonological strings which can be represented in a linear model.

In the following subsections, a number of possible representations for phonotactic descriptions are discussed with respect to the adequacy criteria mentioned in Section 5.1. There have been many different generalisation devices used in order to present the phonotactic data of a language: templates, lists of phoneme combinations, or rules in textual format. They are referred to here as *generalisation devices* rather than *formalisms* since many of them are ad hoc and not formally defined. Others although formally defined are not intended to be adequate for computational purposes but were proposed as descriptions in their own right without any requirements as regards computational interpretability and completeness. Those representations which were developed as part of a computational system will be made explicit below. Section 5.3 goes on to propose a segment-based finite linear representation of phonotactics which is computationally tractable. This is extended to a multi-linear event-based representation in Section 5.6 which is used to describe syllable and complex phonetic event structures.

5.2.1 Templates and Matrices

Two related representation mechanisms for phonotactic descriptions are templates and distribution matrices. A template represents generalisations over consonants (C) and vowels (V) providing a canonical form for the syllable in terms of the number of consonants permitted in the onset and coda. Some examples for German monosyllables are shown here. The complete table can be found in Ternes (1987:171):

V	/aɪ/	Ei
CVC	/bal/	Ball
CCCVC	/ʃtʁaːl/	Strahl
CCCVCC	/ʃtʁant/	Strand
VCCCC	/ɛʁnst/	Ernst
CCCVCCCC	/ʃtʁɔlçst/	(du) strolchst

A generalisation over consonants and vowels allows a template of a German monosyllable to be defined as follows:

CCCVCCCC

All the Cs are optional. This template is often represented by its canonical form where the subscripts and superscripts define the minimum and maximum number of consonants respectively:

$$C_0^3 \ V \ C_0^4$$

Such a CV-template only gives an indication of the length but does not give any information on the distribution of consonants and vowels within the syllable substructures; the template structure alone overgeneralises, providing no restrictions as to which consonants are permitted in which position. Such templates thus place very loose constraints on possible syllables of a language.

Two-dimensional distribution matrices are phoneme based tables of initial consonant clusters followed by a vowel (IC x V) and a vowel followed by final consonant (V x FC) clusters which are used in conjunction with templates in order to restrict the combinations of consonants and vowels. The domain for distribution matrices has usually been the word but they will be discussed below with respect to syllables as this does not alter their overall standing. For each instantiation of the templates C_0^3 V- and -V C_0^4, a distribution matrix is constructed. An small section of such a distribution matrix for English (word-) initial CCV- combinations is shown in Table 5.1. The complete table may be found in Gimson (1970:244-255; cf. also Philipp, 1974 for examples of German data). The difference between accidental and systematic gaps is catered for in the matrix description in so far as *most accidental gaps* and but *not all systematic gaps* are described. However, the matrix description does not overgeneralise. The distribution matrix in Table 5.1 describes, for example, the forms /brɪ/ and /blɪ/. There is no restriction on the combinations of information from different matrices, and therefore these could both be combined with the form /ɪk/ from a -VC distribution matrix. However, since the combination /ft/ does not exist in the CCV- distribution matrix, a form /ftɪk/ is not catered

Table 5.1: Example matrix distribution for some English word-initial CCV-combinations

	ɪ	ɛ	æ	ʌ	ɒ	ʊ	ə	iː	ɑː	ɔː	uː	ɜː
pl	+	+	+	+	+		+	+	+	+	+	
pr	+	+	+	+	+		+	+	+	+	+	
tr	+	+	+	+	+		+	+	+	+	+	
kl	+	+	+	+	+		+	+	+	+	+	+
kr	+	+	+	+	+	+	+	+	+	+	+	
bl	+	+	+	+	+		+	+	+	+		+
br	+	+	+	+	+	+	+	+	+	+	+	+
dr	+	+	+	+	+	+	+	+	+	+	+	
gl	+	+	+	+	+		+	+	+	+	+	
gr	+	+	+	+	+	+	+	+	+		+	

for. The distribution matrix can, however, provide a useful source of information for postulating further generalisations over phonotactic combinations based on structural evidence. If certain consonants are always found in the same context with respect to the template, then this is an indication that they belong to a common class. Furthermore, if particular consonants can be seen to function together as a class in all structural contexts except one, then the generalisation can be made, that this is an accidental gap in the lexicon. For example the distribution of /l/ and /ʁ/ in German is almost identical as can be seen in the subsection of the distribution matrix for two-consonant codas as shown in Table 5.2.

Table 5.2: Subsection of a matrix for two consonant codas following /l/ and /ʁ/

	p	t	k	f	s	ʃ	m	n	ŋ	l	ʁ	x	ç
l	+	+	+	+	+	+	+	+					+
ʁ	+	+	+	+	+	+	+	+		+			+

This is true also for the three consonant codas. However, in the four consonant codas, only /ʁ/ occurs in the context -/nst/. A generalisation about accidental gaps would involve describing the cluster /lnst/ as in the nonsense form *kölnst du?* For this generalisation it is necessary to have some very detailed information about the distribution of the phoneme. Although such matrices provide enough information to generalise to *potential* combinations on the basis of structural descriptions this is not generally done. The distribution matrix together with the template may be said to be descriptively adequate. It could be implemented in a computational system as table lookup, but as an external representation it is not very elegant since a separate matrix is required for each type of cluster (i.e. CV-, CCV-, CCCV-,...). The fact that the distribution matrix allows further generalisations to be made, however, points it in the direction of descriptive adequacy. A representation formalism which explicitly takes generalisation on the basis of functional or structural similarity into account is described in Section 5.3.

Another representation for phonotactics which goes further than the templates given above, is what is termed here as the enhanced template. In this case the templates are compiled with respect to classes of phonemes and thus restrictions are introduced as to which class may occur in which position. Whorf (1940:223) discusses the use of structural formulae in linguistic description and presents a detailed template describing the phonotactic pattern of English word structure. The description of Whorf (1940) does not overgeneralise and can provide a basis for the linear descriptions described later in this chapter.

Other enhanced templates have been proposed which do not distinguish between accidental and systematic gaps. The following enhanced template for a German monosyllable is proposed in Kohler (1977:178):

$$\left(\left\{\begin{array}{c}(C_{\mathrm{a}})(C_{\mathrm{a}})\left\{\begin{array}{c}C_{\mathrm{a}}\\C_{\mathrm{b}}\end{array}\right\}\\C_{\mathrm{c}}\end{array}\right\}\right)\ V\ (C_{\mathrm{b}})\left(\left\{\begin{array}{c}C_{\mathrm{a}}\\C_{\mathrm{b}}\end{array}\right\}\right)(C_{\mathrm{a}})\left(\left\{\begin{array}{cc}C_{\mathrm{a}}&(+C_{\mathrm{a}})\\+C_{\mathrm{a}}&(C_{\mathrm{a}})\end{array}\right\}\right)$$

It goes further than the generalisation in terms of consonants and vowels in that the consonants (C) are divided into classes: C_a being the plosives and the fricatives, C_b the nasals, the lateral and /R/ and C_c containing /h/. Enhanced templates are not restrictive enough in that, despite these constraints, they also overgeneralise. Additional constraints as to which members of a particular class are excluded in certain positions are often provided as text in statements to the effect that all consonants of the classes C_a and C_b can occur prevocalically except /ŋ/ (cf. for example, Kohler, 1977:179). Although the enhanced template does describe actual forms, there is no restriction on potential forms and thus no principled reason for the inclusion or exclusion of certain forms. This enhanced template as is stands is not suitable for a computational system as it drastically overgeneralises if the restrictions in the text are not incorporated into the implementation in some way.

5.2.2 Graphical Representations

Graphical representations are often constructed with respect to *actual* phoneme combinations although on the basis of their topology it is possible to generalise with respect to *potential* forms. Figure 5.1 presents an example of a graphical representation of a subsection the onset of a German monosyllable; the complete representation may be found in Scholz (1972:135).

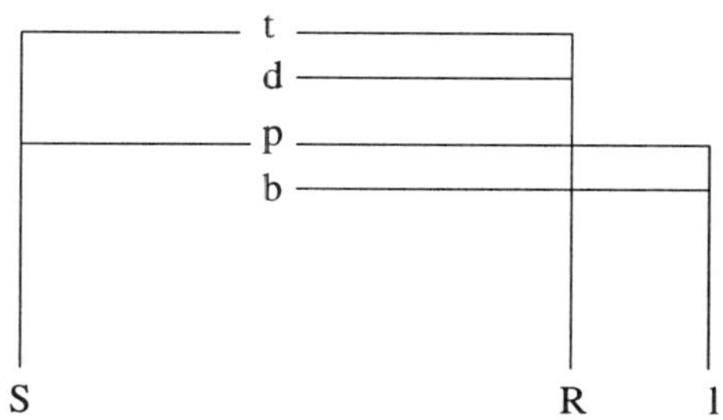

Figure 5.1: Graphical representation of a subsection of the German onset.

This representation could be interpreted as a state labelled finite state automaton by defining the labels in the first two columns as initial states and the labels in the final three columns as final states. The lines connecting them would then represent the arcs of the automaton. However, Scholz does not propose this. He discusses the German monosyllable in considerable detail giving diagrams like the above showing the permissible combinations of phonemes and providing lists of the clusters which occur in the particular positions in terms of a template with restriction classes similar to those given by Kohler (1977). He divides the consonants up into five classes: plosive, fricative, nasal, lateral and vibrant. His constraints are divided into 2 parts: a list of possible combinations of classes and a list of the possible phoneme combinations within these classes. On this basis the graphical representation is to be constructed.

Class combination:	Fricative - Plosive	
Phoneme combinations in this class:	[ʃp]	e.g. [ʃpɪts]
	[ʃt]	e.g. [ʃtaʁk]

Although his description is very detailed, Scholz does not generalise further in his formulation of constraints than the graphical representation in terms of phonemes.

A theory of grammar which makes use of a graphical representation as part of its formalism is Stratificational Grammar (cf. Sampson, 1970). The stratificational approach assumes a strict separation of levels. In stratificational phonology, a phonotactic description traces almost the whole sound system of the language giving full information on the arrangement of phonemes in terms of clusters, syllables, phonological words and higher-rank phonological units. A lower stratum would be concerned with the make-up of phonemes in terms of features but no generalisation is made on this basis in the phonotactic description. An example of a stratificational representation of the onset of an English syllable may be found in Sommerstein (1977:87).

The graphical phonotactic representations presented here are descriptively adequate in that they too cater for accidental gaps but do not offer any principled reason for disregarding systematic gaps. A disadvantage of such graphical representation is thus the lack of further generalisation facilities; due to the reliance on graphs, the already complex topology becomes even more complex if features are introduced.

5.2.3 Feature-based Representations

The idea of using features in phonotactic descriptions is not new. Generative phonology assumed morpheme structure conditions which placed constraints on lexical representations. Feature-based representations of phonotactics in the syllabic domain have reemerged in connection with computational systems. The primary motivation for such representations has been the need for a concise specification using linguistic generalisations in a formalism which can be interpreted by a machine. Two examples of feature-based representations will be presented here.

Context-free phrase-structure rules based on feature sets have been used in allophonic parsing systems (cf. Church, 1987b; Allerhand, 1987) in order to represent a phonotactic grammar whereby the syllable substructures are defined in terms of local hierarchies. Examples of the phrase-structure rules of Church (1987b:97) are presented here:

utterance → syllable*
syllable → onset rhyme
rhyme → peak coda

homorganic-nasal-cluster → labial-nasal labial-obstruent
homorganic-nasal-cluster → coronal-nasal coronal-obstruent
homorganic-nasal-cluster → velar-nasal velar-obstruent

For Church, the phrase-structure rules provide a formal mechanism which has a clear declarative and procedural interpretation, for representing a level between the segment and the lexical entry. However, as discussed in Chapter 3, the local hierarchies of syllable structure are right-branching or linear, a constraint which is not made explicit with context-free phrase-structure rules.

A default-based formalism for the representation of phonotactic constraints which incorporates the notion of linearity explicitly, is the *on-line feature specification model for English syllables* in Figure 5.2 which is taken from Gibbon (1985). This representation consists of a syllable template constructed according to allowable combinations of segments as a discrimination network. The network is constructed by chunking or generalising over syllable substructures and is based on the principle that the number of potential syllables in a language is finite. The model works with underspecification and defaults. The transition labels of the discrimination network are underspecified feature bundles representing natural classes of segments. At particular positions in the syllable network, feature defaults are specified and a convention adopted, namely that once a feature has been specified on a particular transition it remains until it is explicitly altered on a subsequent transition. Thus, vowel harmony may be described by such a network whereby the vowel specifications would remain for subsequent transitions since they would not be relevant for intervening consonants. A finite state formulation can be provided for this model by assuming separate tapes for defaults and the underspecified feature bundles (Gibbon, pers. comm.) together with an elementary nonmonotonic unification operation.

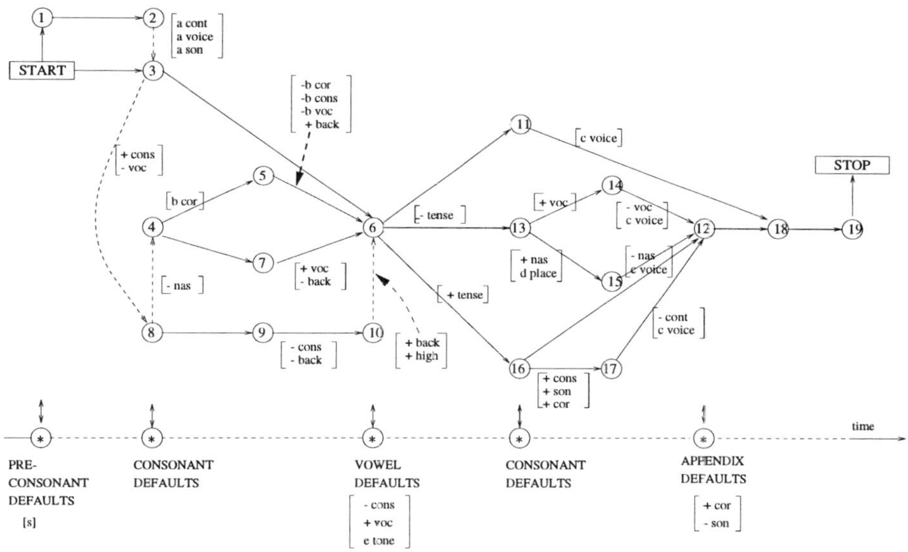

Figure 5.2: The on-line feature specification model for English syllables

5.3 SEGMENT-BASED NETWORK REPRESENTATIONS

In this section, network representation formalisms for phonotactic description will be presented, starting with a simple phoneme-based model and generalising this to a segmental feature-based model. This formalism is extended in Section 5.5 to cater for autosegmental temporal events. All these models consist of networks describing phonotactic rules of a language, in line with the notions of finiteness, linearity and syllable domain as discussed in Section 5.2. The underlying hierarchical syllable structure which expresses distributional "chunking" in phonological strings is not discussed explicitly in terms of onset, peak and coda. Rather these sub-structures and the phonotactic rules which depend on them are expressed in the topology of the network as represented in a transition diagram (as in Figure 5.3). The simplest case is the construc-

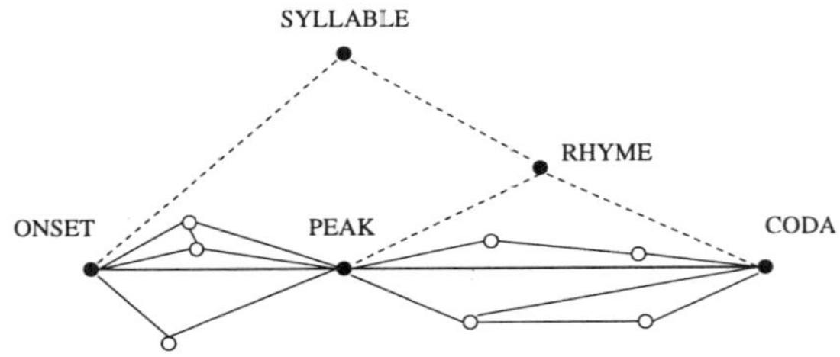

Figure 5.3: Topology of syllable network

tion of a phoneme-based network description using the notion of template and distribution matrix presented above. Figure 5.4 is a graphical representation of a phoneme-based network description showing the permissible consonant combinations in the German syllable onset.

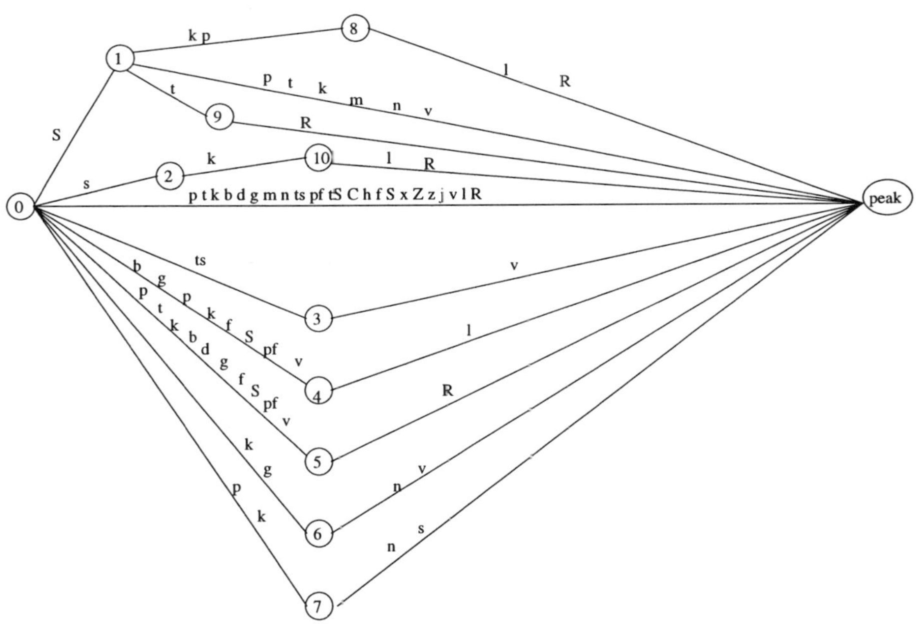

Figure 5.4: Phoneme-based network description of the German syllable onset

The arc labels represent phonemic segments and paths through the network represent possible syllable onsets. The onset /SpR/[1], for example, is described by the path *0 - 1 - 8 - peak.* The fact that some arcs have more than one label

[1]Note SAMPA transcription

is an abbreviation convention for describing single arcs between the same two nodes. The formulation in Figure 5.4 is nondeterministic. Although it is possible to provide a deterministic formulation of syllable structure, the network is less compact. As it stands, the nondeterministic network represents clearly the topology of the German syllable onset in terms of three-consonant clusters, two-consonant clusters and single consonants. However, for reasons of processing efficiency in a computational system, a compilation from a nondeterministic to a deterministic network is desirable (cf. Chapter 7).

The *phoneme-based phonotactic network* can be defined as a triple

$$PN = \langle N, P, D \rangle$$

in terms of:

1. a declaration of nodes *N* (including initial and final nodes)
2. a set of phonemes *P*
3. a collection of arc descriptions *D*

These components are defined in more detail as follows:

1. The nodes *N* relevant for the phoneme-based phonotactic network for German are:

   initial node n_0 : 0
   final node n_{end} : end
   other nodes n_i : 1,2,3,4,...,peak, 11, etc.

2. The set of phonemes for a particular language will not be defined further here.

3. The arc descriptions are in turn defined as a triple

   $$d = \langle n_i, p, n_j \rangle$$

   whereby

   $n_i \in N$: the name of the first node of the arc
   $p \in P$: a phoneme
   $n_j \in N$: the name of the next node.

An example triple taken from Figure 5.4 is <0, p, 5>. This representation is not descriptively adequate in that it contains no generalisation devices and in that unanalysed segments are too coarse as basic categories. It is a declarative formulation in that it makes no statement as to how the network is to be interpreted. However, the formulation in Figure 5.4 can be provided with a procedural interpretation directly in terms of a nondeterministic finite automaton (cf. Chapter 2). Assuming that the arcs are described by a function f, an interpretation is defined in terms of a tuple

$$I = \langle N, P, f, n_0, n_{end} \rangle$$

where

N	: a finite set of nodes	(Q)
P	: a finite input alphabet	($\sum$)
f	: a function mapping N $\times$ P into $\wp$(N)	(δ)
n_0	: the initial node of the network	(q_0)
n_{end}	: the final node of the network	(F)

The symbols given in parentheses are those relating to finite automata in Chapter 2. The function f is interpreted as f:$(n_i,p) \rightarrow \{m_1,\ldots\ldots,m_k\}$ where $m_i \in N$.

5.4 FEATURE-BASED NETWORK REPRESENTATIONS

The feature-based phonotactic network (Carson, 1988) is an extension of the phoneme-based formalism to incorporate linguistically significant generalisations by the use of generalisation devices. Each arc in the feature-based phonotactic network models a phonological segment. The arc descriptions of the network consist of feature bundles containing category features (C-features). Category features are those features required to generalise over a set of phonological symbols. They resemble the classification in terms of N and V features in syntax which, allow generalisation over the conventional syntactic categories of N, V, A and P. Examples are provided below. The advantage of the feature bundle representation over a full phonemic representation is that if distributionally related segments can be viewed in terms of natural classes, then the network is simplified considerably.

The *feature-based phonotactic network* is defined as a quintuple

$$FPN = \langle N, P, F, G, D \rangle$$

of

1. a declaration of nodes N (including initial and final nodes)
2. a phonological feature set P
3. a set F of feature bundles
4. a set of generalisation devices G
5. a collection of arc descriptions D

These components are defined in more detail as follows:

1. The set of nodes *N* relevant for the feature-based phonotactic network for German are:

   initial node n_0 : 0
   final node n_{end} : end
   other nodes n_i : 1, 2, 3, 4, 5, 6, 7, 8, 9, 10, peak, 11, etc.

2. The phonological feature set *P* of attribute-value pairs, is assumed here to be in terms of binary distinctive features based on Chomsky and Halle (1968:301ff). However, any feature system may be used in the representation and the generalisation devices applied accordingly.

3. The set of feature (attribute-value pair) bundles *F* is the set of all *consistent* subsets of *P*. Informally, a feature bundle is termed *consistent* if it does contain any more than one instance of an attribute regardless of its values.

4. Linguistically significant generalisations are expressed in the network by:

   - *natural classes* and *underspecification*: natural classes are defined informally here in terms of feature bundles where the feature bundle describing a natural class of segments contains fewer features than the feature bundle which describes any single element of this class. Thus, the notion of common features can be expressed. Natural classes provide the basis for the classification in terms of category features (C-features) which are those features required by a feature bundle in order to generalise over a set of phonological symbols in some phonotactic position.
   - *variables*: further linguistically significant generalisations are expressed in the network representation by employing Greek-letter variables (e.g. [α voice]), a traditional device of generative phonology. Such variables can range over the values of different features and may also range over adjacent arcs in order to describe phenomena such as assimilation and dissimilation. The restriction of the variables to adjacent arcs means that variables can be used effectively to represent spreading, 'long components' or suprasegmentals.

5. An *arc* in the feature-based phonotactic network is also defined as a triple

$$d = \langle n_i, C_f, n_j \rangle$$

   where

   - $n_i \in N$: the name of the first node of the arc (e.g. "2", "peak", "15")

- $C_f \in F$: a bundle of C-features generalising over the phonological symbols (e.g. the C-feature bundle [-cont, -strid, -voice, -son])
- $n_j \in N$: the name of the next node (e.g. "7", "end")

Possible arcs in a feature-based phonotactic network of the German syllable onset are the following[2]:

<0, [-strid, -cont, -cor], 6> representing {/p/, /b/, /k/, /g/}

<0, [+strid, +cont, -cor, +ant], 6> representing {/f/, /v/}

<0, [+strid, αcont, -voice, αcor, -αant], 6> representing {/S/, /pf/}

<0, [-strid, -cont, +cor], 7> representing {/t/, /d/}

<6, [+cons, +voc], peak> representing {/l/, /r/}

<7, [+cons, +voc, -ant], peak> representing /l/

The path *0 - 6 - peak*, for example, describes the onset combinations /pl/, /pr/, /bl/, /br/, /kl/, /kr/, /gl/ /gr/, /fl/, /fr/, /vl/, /vr/, /Sl/, /Sr/, /pfl/, /pfr/.

Allowing variables to range over adjacent arcs means that the feature-based network representation cannot be mapped directly onto a finite state automaton. In order to cater for this aspect, a type of pushdown automaton with finite control would be required. However, since this is just a generalisation device, a network with variables ranging over adjacent arcs can be compiled into a network structure by multiplying the variable values out onto explicit arcs. In this case, the feature-based phonotactic network FPN can also be interpreted by a nondeterministic finite state automaton analogously to PN.

$$I = \langle N, F, C_f\ g, n_0, n_{end} \rangle$$

where

N	: a finite set of nodes	(Q)
F	: a finite feature bundle input alphabet	(Σ)
C_f	: an element of F	
g	: a function mapping N $\times$ F into $\wp$(N)	(δ)
n_0	: the initial node of the network	(q_0)
n_{end}	: the final node of the network	(F)

The function g is interpreted as g:$(n_i, C_f) \rightarrow \{m_1, \ldots, m_k\}$ where $m_i \in N$. The scanning of C_f is formulated in terms of a subsumption relation on F.

Both the phoneme-based and the feature-based network representations rely on a segmental classification of their basic categories. The motivation for the

[2]Note SAMPA transcription

event concept in phonology as presented in Chapter 4 is that a more flexible phonotactic representation is required as a knowledge component in phonological parsing which incorporates a less rigid segmentation in terms of autonomous events on phonologically relevant tiers, a temporal dimension interpreted in terms of overlap, precedence and inclusion relations between events and which incorporates this into a general notion of underspecification (partial information) as a method of expressing linguistically significant generalisations. This is the prime consideration of Section 5.6.

5.5 FURTHER CONSTRAINTS ON SYLLABLE STRUCTURE

There are other constraints apart from the phonotactic description which should be taken into account when describing syllable structure. The first of these notions concerns the *Sonority Principle*, which dates back to early studies of syllables (for example Jespersen (1904) relies on a segmental classification according to a sonority hierarchy). It is claimed (Goldsmith, 1990:110), that in the onset of the syllable, the segmental material must be arranged in a linear order of increasing sonority from the beginning of the syllable to the nucleus and conversely it must be arranged in the rhyme in a linear order of decreasing sonority from the nucleus. The sonority hierarchy (the inverse of the scale of consonantal strength (e.g. Vennemann, 1988)) classifies phonological segments along the following scale (Figure 5.5):

↑
low vowels
mid vowels
high vowels
glides
liquids
nasals
voiced fricatives
voiceless fricatives
voiced affricates
voiceless affricates
voiced plosives
voiceless plosives

Figure 5.5: The sonority scale

Goldsmith (1990:112) assigns values to the segments classified according to this scale. This notion was originally put forward by Jespersen (1904:186f). Jespersen uses a slightly different classification of the sonority (*Schallfülle*) scale to that of Goldsmith in that he ranks voiceless plosives and voiceless fricatives at the same position. He has a sonority scale ranging form 1 to 8 which allows the graphical representation in Figure 5.6 of the German monosyllable /ʃtʁɔlçst/:

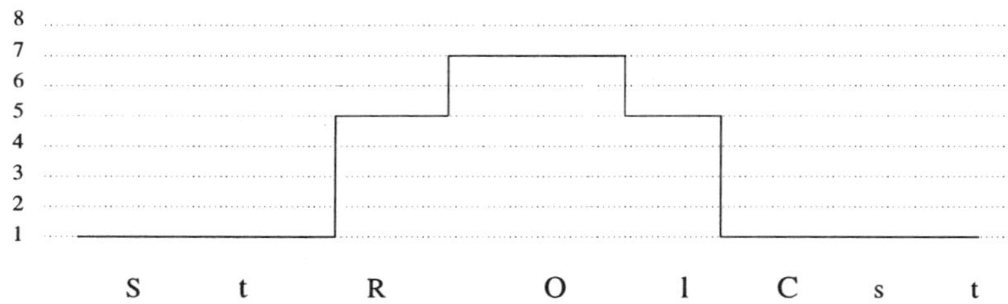

Figure 5.6: Representation of the sonority of the German word *strolchst*

Since a phonotactic network is constructed on the basis of possible combinations of segments, the sonority principle is evident in the final structure. There is a sonority decline from the peak towards the margins. A more recent study which discusses the notion of sonority and suggests a multidimensional approach to segmental strength as parameters of universal phonetic space can be found in Dogil & Luschützky (1990).

A further constraint in connection with syllable structures concerns their type. Syllable structures are defined here with respect to the phonotactic network in terms of *nonreduced* and *reduced* templates. *Nonreduced* and *reduced* syllables depend on the measure of phonological weight which is relevant for the language although they do not correspond directly to the traditional light/heavy distinction (a light syllable having a rhyme which does not branch and a heavy syllable having one that does). An investigation of syllable structures needed for phonological parsing with a computational system showed that for the analysis of phonological words of German, the traditional light/heavy distinction did not provide the correct syllabification and that the following *nonreduced/reduced* structures were more suitable. A *nonreduced* syllable for German, for example, refers to open syllables with long (tense) vowels or syllables with a rhyme consisting of any vowel except schwa and up to four consonants (excluding morphological affixes). A *reduced* syllable caters for open syllables with lax vowels including schwa and syllables with a schwa followed by up to two consonants in the rhyme. This results in the following two syllable types for German:

Nonreduced Syllables:

1. $C_0^3\ V_{\text{tense}}$
2. $C_0^3\ V\ C_1^4$ where V is not schwa

Reduced Syllables:

3. $C_0^2\ V_{\mathrm{lax/schwa}}$
4. $C_0^2\ V_{\mathrm{schwa}}\ C_1^2$

A *nonreduced* syllable may be said to have a preference for stress whereas a *reduced* syllable will always be unstressed (cf. Vennemann (1988)). *Reduced* syllables are only stressable in what are known as ambisyllabic contexts (e.g. /ˈmɪtə/ *Mitte*) where the syllable boundary cannot be determined uniquely. This distinction between syllable types can provide information on stress which can then be further classified on the basis of lexical information (cf. Bleiching, 1991, 1992).

5.6 SYLLABLE EVENT (PHONOTACTIC) NETWORKS

The motivations for the application of the event concept in spoken language recognition were discussed in Chapter 4. In this section, a phonotactic description based on phonological events will be presented which has been used as the phonological knowledge component of a phonological parser of a speech recognition system (cf. Carson-Berndsen 1992a). The representation formalism which has been used is an extension of the network formalisms described in Section 5.3 to cope with multi-linear autosegmental descriptions. It is fundamentally different from other formal devices for autosegmental representation (Kornai, 1991, Bird & Ellison, 1994) in that it models a complete description of German syllable phonotactics within the context of morphological and phonetic components with a full procedural interpretation and an implementation and thus has a different descriptive and operational perspective. It differs from these approaches also in that the descriptive task of the phonotactics is to model all potential syllables using a structured lexicon to distinguish actual from potential forms, and the operational task is to provide well-formedness constraints on phonological and acoustic events in order to restrict the search space of parsers at other levels. This aspect will be discussed in detail in Chapter 7.

On the basis of overlap, precedence and inclusion relations between events, Bird & Klein (1990) define autosegmental representations using the notions of *melody* and *phonological event structure*. The following are their definitions (1990:43):

Definition 5.1 *A melody is an ordered pair* $\tau = < E, \prec^{\circ} >$ *where*

1. *E is a set of events,*
2. $\prec^{\circ}$ *is an irreflexive and an asymmetric relation over E and*
3. *E contains unique maximum and minimum elements with respect to* $\prec$*, denoted* $max(\tau)$ *and* $min(\tau)$ *respectively.*

Definition 5.2 *A phonological event structure is an ordered triple* $\Re = < \mathcal{F}, \prec, ^\circ >$ *where* $F = \{\tau_1, ..., \tau_n\}$ *where each* τ_i *has a distinct type and two sets of ordered pairs specifying the precedence and overlap constraints existing between elements of distinct melodies from* F.

Recall that "$\prec\ ^\circ$" represents immediate precedence, "$\prec$" represents precedence and "$^\circ$" represents overlap.

Bird & Klein (1990) use these notions to describe single utterances (words or syllables) as opposed to a set of syllables or words. This means that for each word or utterance only one phonological event structure is required. This does not pose a problem for fragmental 'hard case' descriptions of language for particular phenomena as is found in much of the literature on autosegmental phonology. However, the domain of phonotactic description is not a single utterance or even a small class of utterances but a description of all *potential* syllables of a language. Adhering strictly to autosegmental theory means that for the purposes of phonotactic description, a disjunction of phonological event structures is required which in turn requires a disjunction of melodies of a single type. The following example indicates the complexity of the problem. The German syllable has the following canonical form:

$$\mathrm{C}_0^3\ \mathrm{V}\ \mathrm{C}_0^5$$

For this structure, melodies would have to be defined for each tier in the phonological representation. Considering the simplest case, the PHONATION tier, where only two properties of the voicing (voiced and voiceless) events can occur, five disjunctive melodies would be required. The notation for event representation corresponds to that defined in Chapter 4.

1. $\tau =< \{< I_1, <\text{PHONATION: voiceless}>>, < I_2, <\text{PHONATION: voiced}>>, < I_3, <\text{PHONATION: voiceless}>>\}, \{< I_1, \prec\ ^\circ, I_2 >, < I_2, \prec\ ^\circ, I_3 >\} >$
2. $\tau =< \{< I_1, <\text{PHONATION: voiceless}>>, < I_2, <\text{PHONATION: voiced}>>, < I_3, <\text{PHONATION: voiceless}>>, < I_4, <\text{PHONATION: voiced}>>, < I_5, <\text{PHONATION: voiceless}>>\}, \{< I_1, \prec\ ^\circ, I_2 >, < I_2, \prec\ ^\circ, I_3 >, < I_3, \prec\ ^\circ, I_4 >, < I_4, \prec\ ^\circ, I_5 >\} >$
3. $\tau =< \{< I_1, <\text{PHONATION: voiced}>>, < I_2, <\text{PHONATION: voiceless}>>, < I_3, <\text{PHONATION: voiced}>>, < I_4, <\text{PHONATION: voiceless}>>\}, \{< I_1, \prec\ ^\circ, I_2 >, < I_2, \prec\ ^\circ, I_3 >, < I_3, \prec\ ^\circ, I_4 >\} >$
4. $\tau =< \{< I_1, <\text{PHONATION: voiced}>>, < I_2, <\text{PHONATION: voiceless}>>\}, \{< I_1, \prec\ ^\circ, I_2 >\} >$
5. $\tau =< \{< I_1, <\text{PHONATION: voiced}>>\}, \{\} >$

Due to the number of event properties (cf. Chapter 4) which can occur on the tiers and to the fact that they are relevant for both consonants and vowels, the definition of melodies for MANNER and PLACE, for example, are much more complicated. Furthermore, in addition to the definition of melodies, the phonological event structure requires the specification of the overlap and precedence constraints existing between elements of distinct melodies. In order to avoid defining a phonological event structure for each possible (i.e. *actual* or *potential*) syllable of a language, a means of generalising over such structures is necessary.

In the rest of this section, a network representation formalism for an autosegmental event-based phonotactic network is presented. Using this representation formalism, complex phonological events such as syllable events, syllable-onset events and syllable-coda events are described. Since the description corresponds to a linguistic *type*, the temporal domain of the description is the relative-time domain (T_{rel}) and the knowledge domain is $K_{phonological}$.

For the purposes of spoken language recognition, a basic distinction should be made between consonantal and vocalic events, the latter being reliably detectable by an acoustic front-end. For this reason the multi-linear autosegmental phonotactic description is defined on the basis of what is termed a *primary tier* which is, in the case of syllable structure, the CV or skeletal tier (cf. Clements & Keyser, 1983). The choice of *primary tier* depends on the application of the description. In fact any tier could be chosen to be the *primary tier* provided its melody is definable in the syllable domain (e.g. sonority tier).

The basic structure of an event-based phonotactic network corresponds to the feature-based phonotactic network discussed in the previous section. However, in contrast to the C-features of the feature-based phonotactic network, the event-based phonotactic network specifies constraints on temporal relations between category events (C-events). Category-events describe a natural class of events with respect to a primary tier. The definition of natural classes of events can be understood analogously to the natural classes of phonological features. The temporal relations which constrain the autosegmental phonotactic description are *overlap* ($\circ$) and *immediate precedence* ($\prec^{\circ}$). Immediate precedence is a constraint which applies along tiers, that is to say, the precedence relations between members of different tiers are not required for the phonotactics.

An *event-based phonotactic network* is defined as a tuple

$$\text{EPN} = \langle \text{N}, \mathcal{E}_{\text{phonological}}, \text{G}, \text{D} \rangle$$

in terms of:

1. a declaration of nodes N (including statement of initial and final nodes)
2. a simplex phonological event set $\mathcal{E}_{\text{phonological}}$
3. a set of generalisation devices G
4. a collection of arc descriptions D

These components are defined in more detail as follows:

1. The declaration of nodes N relevant for the event-based phonotactic network for German are:

   initial node n_0 : 0
   final node n_{end} : end
   other nodes n_i : 1, 2, 3, 4, 5, 6, 7, 8, 9, 10, peak, 11, etc.

2. The simplex phonological event set $\mathcal{E}_{phonological}$ was defined in Chapter 4 (Section 4.4.2) on the basis of the Table 4.4.

3. Generalisations are expressed in the network by:

   - *underspecification*: the classification of phonological segments of the language is in terms of partial information describing natural classes of sounds with structural similarity with respect to syllable structure.
   - *variables*: variables (represented here by α, β, γ etc.) are used in the sense of Greek-letter variables in phonology in order to abbreviate event descriptions (e.g. $< I^1_{pl}, <\text{PLACE: } \alpha >>\}|\alpha \in \{\text{labial, velar}\}$)
   - *pseudo-events*: in order to reduce the size of the phonotactic network, generalisations over arcs have been undertaken by introducing the notion of a pseudo-event. A single phonological event may be referred to (i.e. from the point of view of the primary tier) on two adjacent arcs. For example, both the arc from 0 to peak and the arc from peak to end may refer to a relationship in which an event on the PHONATION tier with the property *voiced* participates; this may be a single voiced event which is consistent throughout the some syllable substructure. Pseudo-events are purely a notational convention in order to generalise over immediate precedence relations.

4. An *arc* in the phonotactic network models an event of the primary tier and is defined as a quintuple

   $$d = < n_i, C_{\mathcal{E}}, \circ, \prec \circ, n_j >$$

   where

   - $n_i \in N$: the name of the first node of the arc (e.g. "2", "peak", "15")
   - $C_{\mathcal{E}} \in \mathcal{E}_{phonological}$: set of C-Events relevant for the arc (e.g. $\{< I^1_{ph}, <\text{PHONATION: voiceless}>>, < I^1_{ma}, <\text{MANNER: plosive}>>, < I^1_{pl}, <\text{PLACE: } \alpha >>\}|\alpha \in \{\text{labial, velar}\}$. The sub/superscripts are used to indicate unique intervals in T_{rel} with respect to tiers and arcs. The variable α is specified together with the set it abbreviates.

- $\circ$: set of overlap constraints expressing relations (e.g. $< I^1_{ma}, \circ, I^1_{pl} >\}$ which states an overlap constraint between an event on the MANNER tier and an event on the PLACE tier in the intervals I^1_{ma} and I^1_{pl}
- $\prec \circ$: set of immediate precedence constraints expressing relations (e.g. $\{< I^6_{ma}, \prec \circ, I^7_{ma} >\}$ which states an immediate precedence constraint between events on the MANNER tier in the intervals I^6_{ma} and I^7_{ma})
- $n_j \in N$: name of next node (e.g. "7", "end")

A selection of arc descriptions for one path through event-based phonotactic network for German are given below. Due to its length, the complete event-based phonotactic description is provided in appendix C.

<0,
$\{< I^0_{ph}$, <PHONATION: voiceless>>, $< I^0_{ma}$, <MANNER: fricative>>, $< I^0_{lp}$, <PLACE: palato-alveolar>>}
$\{< I^0_{ph}, \circ, I^0_{ma} >, < I^0_{ma}, \circ, I^0_{pl} >, < I^0_{ph}, \circ, I^0_{pl} >\}$,
$\{< I^0_{ph}, \prec \circ, I^1_{ph} >, < I^0_{ma}, \prec \circ, I^1_{ma} >, < I^0_{pl}, \prec \circ, I^1_{pl} >\}$,
1>

<1,
$\{< I^1_{ph}$, <PHONATION: voiceless>>, $< I^1_{ma}$, <MANNER: plosive>>, $< I^1_{pl}$, <PLACE: α >>}$|\alpha \in$ {labial, velar},
$\{< I^1_{ph}, \circ, I^1_{ma} >, < I^1_{ma}, \circ, I^1_{pl} >, < I_{ph1}, \circ, I^1_{pl} >\}$,
$\{< I^1_{ph}, \prec \circ, I^8_{ph} >, < I^1_{ma}, \prec \circ, I^8_{ma} >, < I^1_{pl}, \prec \circ, I^8_{pl} >\}$,
8>

<8,
$\{< I^8_{ph}$, <PHONATION: voiced>>, $< I^8_{ma}$, <MANNER: lateral>>},
$\{< I^8_{ph}, \circ, I^8_{ma} >\}$,
$\{< I^8_{ma}, \prec \circ, I^{peak}_{ma} >\}$,
peak>

<peak,
{<$< I^{peak}_{ma}$, <MANNER: vowellike>>, $< I^{peak}_{pl}$, <PLACE: α >>}$|\alpha \in$ {front, back},
$\{< I^{peak}_{ma}, \circ, I^{peak}_{pl} >\}$,
$\{< I^{peak}_{ma}, \prec \circ, I^{11}_{ma} >, < I^{peak}_{pl}, \prec \circ, I^{11}_{pl} >\}$,
11>

<11,
$\{< I^{11}_{ph}$, <PHONATION: voiceless>>, $< I^{11}_{ma}$, <MANNER: affricate>>, $< I^{11}_{pl}$, <PLACE: α >>}$|\alpha \in$ {apical, palato-alveolar} ,
$\{< I^{11}_{ph}, \circ, I^{11}_{ma} >, < I^{11}_{ma}, \circ, I^{11}_{pl} >, < I^{11}_{ph}, \circ, I^{11}_{pl} >\}$,
{},
end>

The subscripts refer to the tiers (*ph*onation, *ma*nner, *pl*ace) and the superscripts refer to the nodes of the arcs.

A transition diagram showing the overall structure of a German syllable is presented in Figure 5.7. In this figure, the phonological segments which can be classified by constraints on temporal relations have been grouped together in square brackets [ ]. For example, on the arc from 0 to 6

$$\begin{bmatrix} k \\ g \end{bmatrix} \textit{represents} \begin{bmatrix} < I^0_{\mathrm{ma}}, < \textsc{manner} : \textit{plosive} >> \\ \circ \\ < I^0_{\mathrm{pl}}, < \textsc{place} : \textit{velar} >> \end{bmatrix} \tag{5.1}$$

The topology of the diagram describes the primary tier of a syllable event.

The interpretation of the event-based phonotactic network EPN as an automaton is given as a tuple

$$\mathrm{I} = < N, \mathcal{E}_{\mathrm{phonological}}, \circ, \prec \circ, \mathrm{R}, h, \mathrm{n}_0, \mathrm{n}_{\mathrm{end}} >$$

where

N	: a finite set of nodes	(Q)
$\mathcal{E}_{\mathrm{phonological}}$	: a finite input alphabet of phonological events	
$\circ$	: the set of all possible overlap relations in $\mathcal{E}_{\mathrm{phonological}}$	
$\prec \circ$	: the set of all possible immediate precedence relations in $\mathcal{E}_{\mathrm{phonological}}$	
R	: $\circ \cup \prec \circ$ a finite input alphabet of event relations on $\mathcal{E}_{\mathrm{phonological}}$	($\sum$)
h	: a function mapping N $\times$ R into $\wp$(N)	(δ)
n_0	: the initial node of the network	(q_0)
$\mathrm{n}_{\mathrm{end}}$	: the final node of the network	(F)

The function h is interpreted as h:($\mathrm{n_i}$,r) $\rightarrow$ {m_1,.....,$\mathrm{m_k}$} where $\mathrm{m_i} \in$ N. The scanning of r $\in$ R is discussed in detail in Chapter 7 where this procedural interpretation will serve as a basis for the operational semantics of the event processor for phonological parsing. In this way, the event-based phonotactic network can be used to recognise syllable events (defined by all possible paths between the nodes *0* and *end*), syllable-part events such as syllable-onset events (defined by all possible paths between the nodes *0* and *peak*) and syllable-type events. A distinction is made between two syllable types for German: nonreduced syllables and reduced syllables (as defined in Section 5.5). For each syllable type an event-based phonotactic network has been constructed. This distinction between syllable-type events can provide information on stress which can be further classified on the basis of lexical information (cf. Bleiching, 1991, 1992).

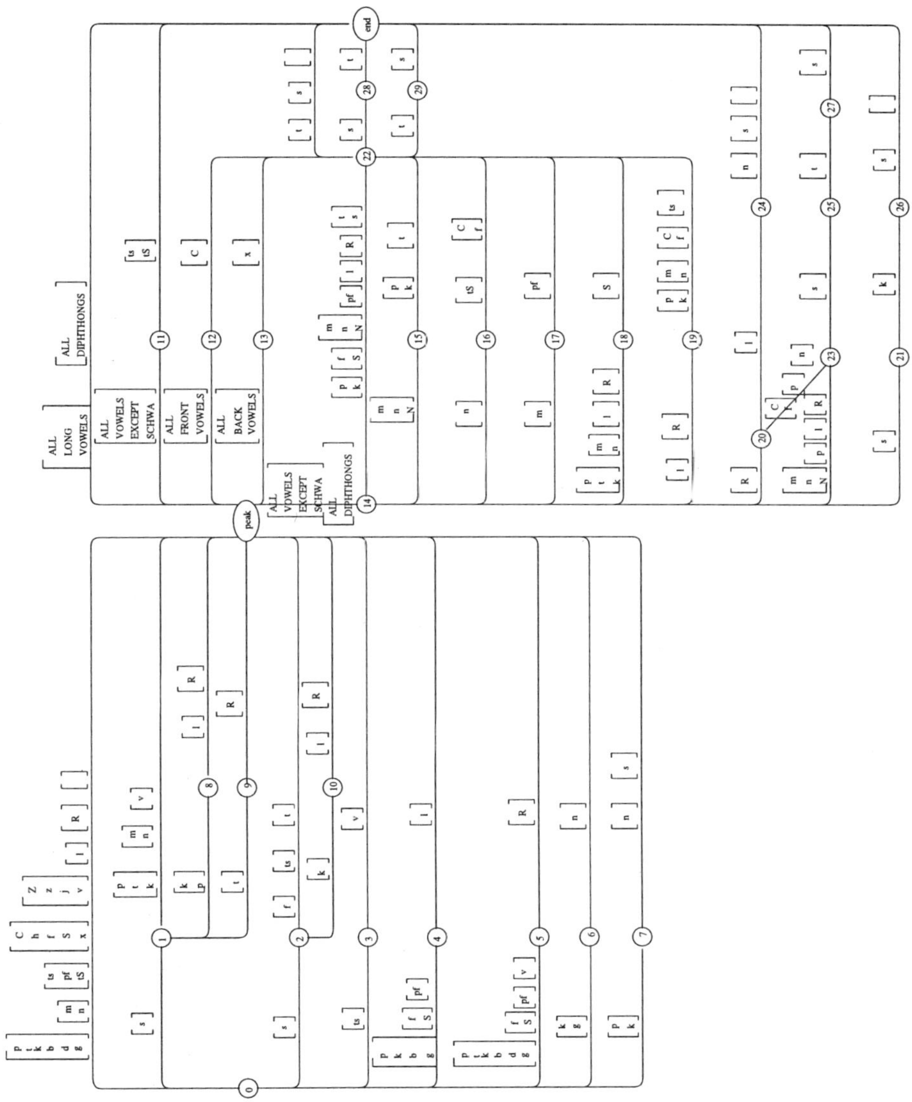

Figure 5.7: German phonotactic network

This section has been concerned with the definition of a representation formalism for an event-based description of syllable structure. It was shown how, in this formalism, a complete event-based phonotactics of German syllable events can be described within the temporal domain T_{rel} and the knowledge domain $K_{phonological}$. The network representation formalism for event descriptions can also be utilised in the phonetic knowledge domain. This is the main concern of the next section.

5.7 COMPLEX PHONETIC EVENT NETWORKS

In Chapter 4, simplex and complex events were defined with respect to some knowledge domain K. A complex event in one knowledge domain may be a simplex event in a higher knowledge domain. The event-based phonotactic network described in Section 5.6 uses simplex events in the knowledge domain $K_{phonological}$ as its primitives and on the basis of temporal relations between these simplex events, complex phonological events are described compositionally in the network. These simplex phonological events are complex phonetic events in the knowledge domain $K_{phonetic}$ which are described compositionally on the basis of temporal relations between simplex phonetic events. The rest of this chapter is concerned with descriptions of complex phonetic events.

For each complex phonetic event in $K_{phonetic}$ there exists an event-based network description analogous to the descriptions in $K_{phonological}$ for complex phonological events such as syllable events. The complex phonetic event networks differ from the networks describing syllable structure in that they are based on the temporal relations overlap and precedence. Temporal inclusion is not catered for in the current descriptions but may provide some important cues for a later version. Immediate precedence is not required for descriptions of complex phonetic events since each tier has currently only one relevant property (cf. Section 4.4.1).

An *complex phonetic event network* is defined as a tuple

$$\text{CPN} = <\text{N}, \mathcal{E}_{\text{phonetic}}, \text{D}>$$

in terms of:

1. a declaration of nodes N (including statement of initial and final nodes)
2. a simplex phonetic event set $\mathcal{E}_{phonetic}$
3. a collection of arc descriptions D

These components are defined in more detail as follows:

1. The set of nodes N relevant for the complex phonetic event network for German are:

initial node n_0 : 0
final node n_{end} : end
other nodes n_i : 1, 2, 3, 4, 5, etc

2. The simplex phonetic event set $\mathcal{E}_{phonetic}$ was defined in Chapter 4 (Section 4.4.1) on the basis of the Table 4.3.

3. An *arc* in the complex phonetic event network is defined as a quintuple

$$d = < n_i, C_{\mathcal{E}}, {}^{\circ}, \prec, n_j >$$

where

- $n_i \in N$: the name of the first node of the arc (e.g. "2", "3")
- $C_{\mathcal{E}} \in \mathcal{E}_{phonetic}$: set of C-Events relevant for the arc (e.g. $< I^1_{oc}, \text{<occlusion>}>, < I^1_{re}, \text{<release>}>\}$. The sub/superscripts are used to indicate unique intervals in T_{rel} with respect to tiers and arcs.
- $^{\circ}$: set of overlap constraints expressing relations (e.g. $< I^1_{oc}, {}^{\circ}, I^1_{re} >\}$ which states an overlap constraint between an occlusion event and a release event in the intervals I^1_{oc} and I^1_{re})
- $\prec {}^{\circ}$: set of immediate precedence constraints expressing relations (e.g. $\{< I^6_{re}, \prec, I^7_{no} >\}$ which states an immediate precedence constraint between an occlusion event and a noise event in the intervals I^6_{re} and I^7_{no})
- $n_j \in N$: name of next node (e.g. "7", "end")

An example complex phonetic event network describing the event *plosive* is presented here.

<0,
$\{< I^0_{oc}, \text{<occlusion>}>\}$
$\{\}$,
$\{< I^0_{oc}, \prec, I^1_{re} >\}$,
1>

<0,
$\{< I^0_{oc}, \text{<occlusion>}>, < I^0_{re}, \text{<release>}>\}$
$\{< I^0_{oc}, {}^{\circ}, I^0_{re} >\}$,
$\{< I^0_{oc}, \prec, I^2_{fr} >, < I^0_{re}, \prec, I^2_{fr} >\}$,
2>

<0,
$\{< I^0_{oc}, \text{<occlusion>}>, < I^0_{re}, \text{<release>}>\}$
$\{< I^0_{oc}, {}^{\circ}, I^0_{re} >\}$,
$\{\}$,
end>

<1,
{< I^1_{re}, <release>>}
{},
{< $I^1_{\mathrm{re}}, \prec, I^2_{\mathrm{fr}}$ >},
2>

<1,
{< I^1_{re}, <release>>}
{},
{},
end>

<1,
{< I^1_{re}, <release>>, < I^1_{fr}, <frication>}
{< $I^1_{\mathrm{re}}, \circ, I^1_{\mathrm{fr}}$ >},
{},
end>

<2,
{< I^2_{fr}, <frication>}
{},
{},
end>

The complex phonetic event *plosive* is defined by all possible paths from node *0* to node *end*. Other complex phonetic events are defined analogously.

Phonotactic and complex event descriptions have been presented in Sections 5.6 and 5.7 which go some way towards fulfilling explanatory adequacy criterion mentioned in 5.1. The descriptions explicitly take the notions of finiteness, linearity, and domain of application into account and have been modelled using formalisms which have a declarative and procedural interpretation. The fact that this procedural interpretation is invertible means that the complex phonetic and phonological event descriptions have applications in both synthesis (generation) and recognition (analysis). The phonotactic networks act as flexible knowledge components which cater not only for *actual* syllables of the language but also generalises to all *potential* syllables. This is the type of linguistic information is needed below the word level in order to be able to cater for neologisms which are not found in the lexicon. The event-based network representation allows for a multi-linear description of syllables. The complex phonetic networks describe possible autosegments on individual phonological tiers.

It will be seen in Chapter 7 that the event-based network descriptions of phonotactics and complex phonetic events can be used in a phonological parser in order to enhance underspecified phonetic input on the basis of structural regularities and it is this feature which distinguishes event-based phonological parsing from other approaches to spoken language recognition.

6

EXCURSUS: CONSTRAINT-BASED SEGMENTAL PHONOLOGICAL PARSING

This chapter has been termed an excursus because it describes a segmental approach to phonological parsing which uses only a few of the notions introduced in the previous chapters. However, this segmental approach defines the starting point of constraint-based phonological parsing by employing segment-based phonotactic structure constraints in a feature-based phonotactic network (FPN) to provide strings of allophonic feature bundles with a phonological representation in terms of syllables. The FPN has been extended to represent an allophonic and a phonemic level and is interpreted by a finite state transducer. The nonsegmental approach which relies on events is presented in the next chapter.

Phonological parsing was introduced in Chapter 1 as the analysis of a temporal stream of phonetic information on the basis of phonological knowledge about a language; extra-lexically it assigns phonetic input to a structure which is relevant at the phonological level. This process is usually regarded as the abstraction of invariant, language-specific phonological properties from the variant information contained in the phonetic representation; it should not be confused, however, with the less conventional meaning found in generative phonologies of application of phonological rules, as described in Chapter 2. In phonological parsing, it is the phonetic realisation which acts as the starting point and the aim is to find a suitable phonological interpretation.

Much of the work in phonology tends to be restricted to the analysis of particular problems among "ideal" standard data, that is to say, with prototypical data containing phenomena which are assumed by phonologists to appear in classified and segmented spoken language. "Real" speech directly observed temporal data are, however, filled with ungrammaticalities and what is termed by the speech technology community as noise. It is, therefore, essential that computational phonology concerns itself, not only with being able to provide such knowledge bases and processors as can handle standard forms, but also such as can handle the phenomena of continuous fast speech.

In this chapter and the next, two different constraint-based approaches to phonological parsing are described which use aspects of the formalisms discussed in the preceding chapters. In each case concrete phonological parsers are presented which, although they have many similarities, are conceptually different in that the first is feature-based or segmental and the second is nonsegmental multi-linear event-based. The feature-based segmental parser (PHOPA) is designed from the point of view of computational phonology as a discipline in its own right and thus incorporates notions from traditional phonology and computational linguistics, the prime interest being questions of modelling, of generalisation and representation. The event-based parser used in connection with *Time Map Phonology* which is described in Chapter 7 is more spoken language oriented in that it was designed as part of a speech recognition system and is therefore directed at solving problems which are to be found in "real" phonetic data.

In the first sections of this chapter a constraint-based phonological parser, PHOPA, is introduced which uses as its knowledge base an extension of the feature-based phonotactic network described in Chapter 5 in order to parse a phonetic sequence into phonological words. The PHOPA model is described with respect to its declarative and the procedural domains. The representation domain of PHOPA refers to the representations at the phonetic (or allophonic) and phonological levels, and the processing domain refers to the processor, a finite state transducer which is used to interpret these representations. In Section 6.4, a model for the segmental processing of speech variants is introduced which provides a first approximation to a solution to the projection problem in spoken language recognition. This model is coupled with a word network generator which compiles specialised word models for verification purposes.

6.1 THE PHOPA MODEL: AN OVERVIEW

The PHOPA model of phonological analysis (Carson, 1988; Carson-Berndsen et. al, 1989) contains a fully implemented segmental parser for phonological words of German. PHOPA uses as its linguistic knowledge base a declarative system of constraints on the allophonic syntax and of the phonotactics of a particular language, in this case German, in order to parse a phonetic sequence into phonological words. The PHOPA system has two declarative feature-based linguistic knowledge components: a feature-based phonotactic network and a redundancy rule component. The feature-based phonotactic network, which is not restricted to a particular feature set, in principle, represents the syntagmatic constraints on the allophonic and phonological domains. It is an extension of the feature-based phonotactic network described in Chapter 5 in that it describes two levels in a single network. It serves as a phonological word template for German consisting of *nonreduced* and *reduced* syllables. The arcs in the feature-based phonotactic network represent transduction relations which rely on a restricted form of unification and elementary operations over sets in order to express the correspondences between the allophonic and the phonological domains. The redundancy rule component is a system of language-specific and universal paradigmatic constraints on the allophonic and phonological domains expressed in terms of feature cooccurrence restrictions and feature specification defaults. These phonological rule systems will be described in detail in Section 6.2. An overview of the architecture of the system is presented in Figure 6.1.

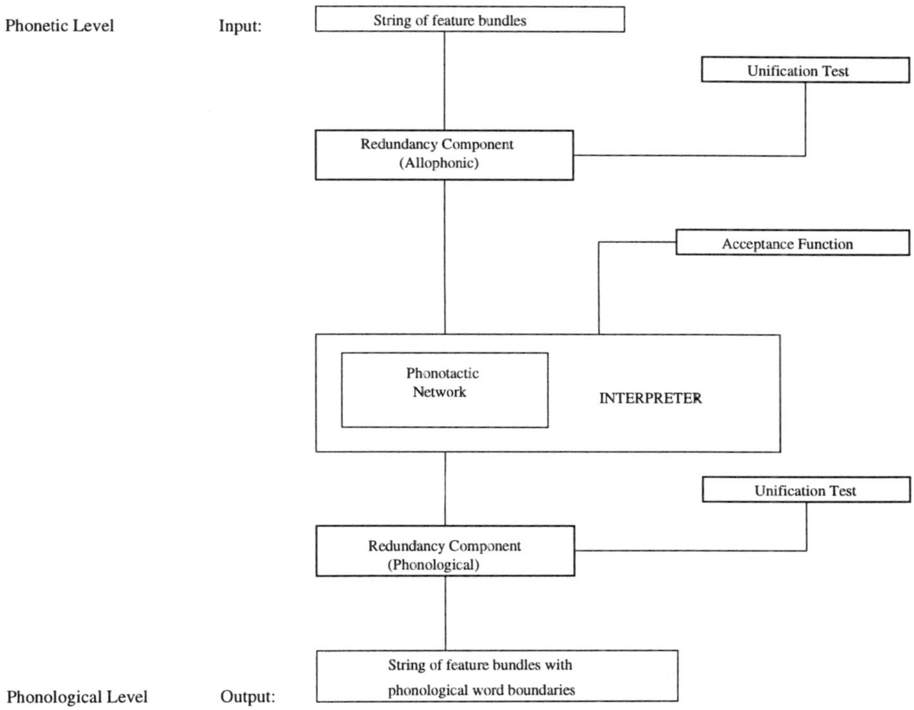

Figure 6.1: Overview of PhoPa

The architecture of the PhoPa system is not language-dependent; that is to say, the linguistic knowledge components and the feature sets can be substituted according to the language which is to be analysed. Similarly any feature classification (unary, binary, multivalued) may in principle be chosen as long as the feature-based phonotactic network and redundancy rules are compiled accordingly. In this discussion, a binary feature set for German, based on Chomsky and Halle (1968), is taken for the phonological set and adapted to incorporate allophonic information. There are thus two sets of features, one for the allophonic domain and one for the phonological domain. The interpreter used in the PhoPa system is a nondeterministic finite-state transducer but since the representation is independent of the processing strategy, any interpreter (chart parser etc.) could be substituted here.

The phonetic input to PhoPa is a finite, in general, well-formed string of allophonic feature bundle segments and the phonological interpretation is a linear representation of phonological words of German. It is obvious, therefore, that this parser is designed to cater only for what is termed labelled ("ideal") data. It assumes that an acoustic front-end processor has been able to classify over allophonic feature bundle segments. This assumption is too unrealistic and will be modified for real data in Chapter 7. However, the motivation for PhoPa lies in its treatment of underspecified phonetic input using a thorough description of relevant phonological structures. The parser described here should be regarded as the starting point, from a computational phonological point of view, for the development of a robust parser which can be used in a real speech recognition system. It will become clear that since this segmental model incorporates no explicit concept of temporality, it is more suited to descriptive computational phonological studies such as the verification of phonological descriptions than to integration into a spoken language recognition system.

PhoPa is now discussed with respect to the declarative domain in Section 6.2 and with respect to the procedural domain in Section 6.3.

6.2 THE DECLARATIVE DOMAIN OF PHOPA

The interest in computational systems for phonology has led away from the derivational or procedural aspect of phonological description towards a declarative constraint-based non-derivational approach (cf. Scobbie, (1991a)). Thus, rather than writing rules which allow the derivation of a surface form from some abstract underlying representation, declarative phonology is concerned with writing constraints on well-formedness of phonological representations. In this section, the main concern is the constraint formalism used in the phonological linguistic knowledge components of the feature-based segmental parser. The PhoPa model uses the notion of a feature-based network and feature bundle representations for its knowledge components together with generalisation devices of feature-based segmental phonology such as underspecification, redundancy, natural classes and Greek-letter variables. Underspecification and redundancy are considered from a new angle in phonological parsing and for this reason they are discussed separately in the next subsections.

6.2.1 Redundancy and Underspecification

Both in the sense of Universal Grammar and language-specifically, there are implications which exist between features in particular contexts. The main concern of Underspecification Theory is the minimisation of redundancy in feature specifications thus providing a suitable evaluation metric for phonological representations (cf. Halle, 1959). The notion has existed since the work of Chomsky & Halle (1968) where it was considered desirable to have underspecified feature structures in the underlying lexical representation. It was the predictable values which were not required to be specified in these representations as they could be provided by rule in order to generate the fully specified phonetic realisations. Further work on this problem has led this type of underspecification to be known as segment-based *contrastive specification*. Archangeli (1988) in her overview of recent work in the area of underspecification in phonology proposes a theory of Underspecification Phonology which uses another type of underspecification known as *radical underspecification*, with features rather than segments as the phonological primitives. This type of underspecification goes further than contrastive specification by removing all redundancies allowing for a variety of representations providing a basis for the notion of preferability or *markedness* in Universal Grammar. Underspecification phonology differentiates between language-specific and universal defaults. The difference between these two types can be formalised using the Generalised Phrase Structure Grammar notions of *feature cooccurrence restrictions* and *feature specification defaults* (as also done by Calder & Bird, 1991). Contrastive specification uses only feature cooccurrence restrictions and radical underspecification uses both of these devices.

Until now the discussion has been concentrated on underspecification for the purposes of minimising redundancy in the underlying phonological representations. It has been assumed so far that this is only relevant at the phonological level and that the phonetic realisations or surface forms will always be fully specified. Keating (1988) provides a detailed report on cases where the phonetic realisation is seen to remain underspecified for certain features. Her results, based on data from coarticulation, show that underspecified representations are needed at the phonetic level in order to account for particular phenomena. Her discussion leads her to posit that such information would be useful to a language learner for the purpose of distinguishing which features are specified in the phonetic realisation of a phonological segment. It is this point which is directly connected to the problems of speech recognition in the sense that it refers to analysis rather than generation.

In the context of phonological parsing in the PhoPa model, underspecification is applied in two separate areas: phonology and phonetics. Firstly *top-down*, in the more traditional sense of wanting to provide a precise, economic representation of allophonic and phonological well-formedness, the declarative knowledge is constructed in terms of networks of underspecified feature bundles. *Top-down* underspecification provides a way of expressing linguistically significant generalisations over this data. Secondly *bottom-up*, since a feature extraction front-end is not able to provide the sort of full specifications required by the phonologists, the phonetic input data to the phonological parser will be underspecified. *Bottom-up* underspecification, in this context, is an inevitable rather than desirable state of affairs. However, work on phonological top-down underspecification can provide further feature specifications, namely those which can be predicted from context. Rather than being applied to underspecified underlying phonological representations, feature cooccurrence restrictions and feature specification defaults can be applied to an underspecified phonetic representation in order to find a more fully specified phonological interpretation. The way in which underspecification is used for phonological parsing is described in connection with phonological rule systems in the next section.

6.2.2 Phonological Rule Systems and Constraints

In this section, the phonological rules used in segmental phonological parsing are discussed as a declarative system of constraints on the syntagmatic and paradigmatic relations of phonological representations. The segmental PhoPa model has two linguistic knowledge components: the feature-based phonotactic network which expresses syntagmatic phonological structure constraints and the redundancy rule component which expresses cooccurrence constraints on features. These will now be defined on the basis of a binary feature set. However, as mentioned above, only the knowledge components of the PhoPa system rely on a feature classification and these components can operate on another feature set if preferred.

The feature-based phonotactic network is a declarative, constraint-based formalism. Depending on the semantics, it can have two possible procedural interpretations. It can be used as a linguistic knowledge component either for a parser or for a generator of possible syllables (or phonological words) of a language. The representation formalism for the feature-based phonotactic network described below is an extension of the network formalism described in Chapter 5. Rather than expressing the allophonic syntax and the phonotactics as two separate representations, the feature-based phonotactic network can be regarded more in the sense of a two-tape finite-state transition network in that it describes a cascade of two network representations describing the constraints on the allophonic syntax and phonotactics of a language. Each arc in the feature-based phonotactic network models a phonological segment. The advantage of the feature bundle representation over a full phonemic representation is that segments can be viewed in terms of natural classes, thus incorporating linguistically significant generalisations, which simplifies the network considerably. The arc descriptions of the network consist of pairs of feature bundles each containing category features (C-features as defined in Chapter 5). One of these feature bundles represents allophonic specifications and the other phonological specifications; both are in general underspecified.

The feature-based phonotactic network of PhoPa is defined analogously to the feature-based phonotactic network presented in Section 5.3 in terms of:

1. a declaration of nodes N (including statement of initial and final nodes)
2. an allophonic feature set F_a
3. a phonological feature set F_p
4. a set of generalisation devices G
5. a collection of arc descriptions D

These components are defined in more detail as follows:

1. The nodes N relevant for the feature-based phonotactic network for German are:

initial node	:	0
final node	:	end
other nodes	:	2, 3, 4, 5, 6, 7, 8, 9, 10, peak, 11, etc.

2. The allophonic feature set F_a, used in this version of the PHOPA system, is defined in terms of binary features. It is an extension of the phonological feature set F_p in that features such as aspiration (+ asp) and syllabicity (+ syll) are included in the set.
3. The phonological feature set F_p, used in this version of the PHOPA system, is defined in terms of binary distinctive features based on Chomsky and Halle (1968).
4. Linguistically significant generalisations are expressed in this network by natural classes, underspecification and variables as defined in Section 5.5.
5. An *arc* in the feature-based phonotactic network is defined as a quadruple:

   $$d = <n_i, C_a, C_p, n_j>$$

   where

   - $n_i \in N$: the name of the first node of the arc (e.g. "2", "peak", "15")
   - $C_a \in F_a$: the C-FEATURES generalising over the allophonic symbols (e.g. the C-FEATURE bundle [-cont, -strid, -voice, -son, +asp])
   - $C_p \in F_p$: the C-FEATURES generalising over the phonological symbols (e.g. the C-FEATURE bundle [-cont, -strid, -voice, -son])
   - $n_j \in N$: the name of the next node (e.g. "7", "end")

A selection of arc descriptions for German syllable onset in the extended feature-based phonotactic network is given here [1]:

<0, [-strid, -cont, -cor, -voice, +asp], [-strid, -cont, -cor, -voice], peak>

representing {/pH/:/p/, /tH/:/t/, /kH/:/k/}

<0, [-strid, -cont, -cor], [-strid, -cont, -cor], 6>

representing {/p/:/p/, /b/:/b/, /k/:/k/, /g/:/g/}

<0, [+strid, +cont, -cor, +ant], [+strid, +cont, -cor, +ant], 6>

representing {/f/:/f/, /v/:/v/}

<0, [+strid, αcont, -voice, αcor, -αant], [+strid, αcont, -voice, αcor, -αant], 6>

representing {/S/:/S/, /pf/:/pf/}

<0, [-strid, -cont, +cor], [-strid, -cont, +cor], 7>

representing {/t/:/t/, /d/:/d/}

<6, [+cons, +voc], [+cons, +voc], peak>

representing {/l/:/l/, /r/:/r/}

<7, [+cons, +voc, -ant], [+cons, +voc, -ant], peak>

representing /l/:/l/

The feature-based phonotactic network is used not only to describe single possible syllables, but also phonological words. The feature-based phonotactic network for German consists of two subnets describing *nonreduced* and *reduced* syllables respectively (cf. Section 5.5). The PhoPa model was initially limited to bisyllabic words and thus, a constraint was imposed such that a phonological word must contain at least one *nonreduced* syllable. Monosyllabic words must be nonreduced syllables. The basic structure is depicted in the graphical representation in Figure 6.2 where NR represents a nonreduced syllable and R represents a reduced syllable. The possible combinations described in Figure 6.2 are bisyllabic words with the structure nonreduced-nonreduced, nonreduced-reduced and reduced-nonreduced. Tests have shown that this structure can be generalised to multi-syllable phonological words.

6.2.3 The Redundancy Rule Component

The redundancy rule component is the linguistic knowledge base in which constraints on the well-formedness of feature-based segmental descriptions are described. These represent the paradigmatic aspect of the phonological descrip-

[1] Note SAMPA transcription

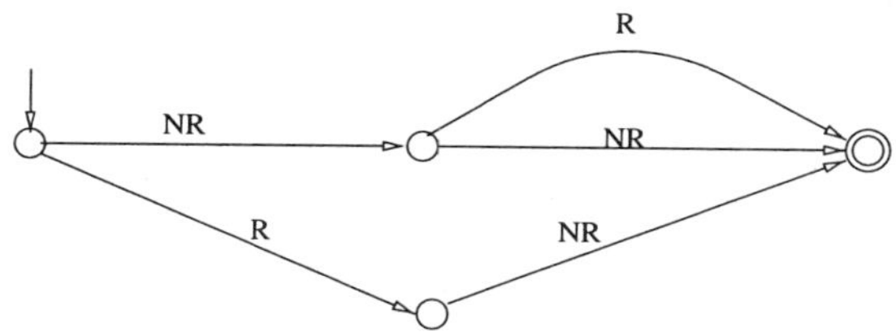

Figure 6.2: Graphical representation of constraints on syllable types in bisyllabic phonological words

tion. This component contains *all* positive implications between the features of a particular set. This is in contrast to the traditional view of redundancy rules in phonology which assume that a standard minimal feature specification exists and that the redundant information is then "filled in" by rule. However, in phonological parsing of underspecified phonetic input, it is not certain which features will have been recognised by the front-end. Thus, for every possible recognisable feature, there must exist a set of feature cooccurrence constraints. There is thus no formal distinction made in this component between language-specific and universal constraints.

The redundancy rule component is divided into two domains: the allophonic feature implications and the phonological feature implications which correspond to the feature cooccurrence restrictions (FCRs) of Generalised Phrase Structure Grammar. The format of the rules from each domain is as follows:

$$\mathrm{X} \in \mathrm{F_a} \Rightarrow \mathrm{Y} \in \mathrm{F_a}$$

where $\mathrm{F_a}$ is the allophonic feature set and the left hand side of the rule specifies the condition for application. Where there are several rules for a particular X, the more general format can be used:

$$\mathrm{FB} \Rightarrow \mathrm{FB'}$$

where both FB and FB' are feature bundles consistent in $\mathrm{F_a}$. This also means that the rules are not required to carry a single feature specification in the left hand side on the rule. The rules are defined analogously for the phonological domain. The following examples express well-formedness constraints on phonological features within segments.

[+ high] ⇒ [- low]

[+ nas] ⇒ [+son, +voice, -voc, +cons, -low]

[+ voc] ⇒ [+son, +voice, -nas]

[+voc, +cons] ⇒ [-high, -low, +cor, +voice, +cont, -nas, -strid, +son]

Radical underspecification in the form of feature specification defaults are more relevant in the lexical domain of description and have not been discussed in this section as they are not considered part of the redundancy component.

6.3 THE PROCEDURAL DOMAIN OF PHOPA

In this section the procedural aspects of the PHOPA model are discussed. Firstly, the overall algorithm is presented and then the application of the redundancy rules and the processing of the allophonic and phonotactic constraints are described in detail. An overview of the architecture of the PHOPA system was shown in Figure 6.1 above. Unification and transduction play a major role in this model. Redundancy rule application, which should be regarded separately from the parsing process, uses unification as its main mechanism. Parsing is carried out by a finite state transducer using a mapping procedure based on unification which is described below. Transduction takes place between the allophonic and the phonological levels.

6.3.1 Parsing with the PHOPA System

The PHOPA parser is phonetic-data-driven (bottom-up) which means that variant information in allophonic specifications which is needed for parsing syllable structures is discarded as it is no longer relevant at the phonological level. The work of the parser is very similar to the procedure proposed by Church (1987b, cf. also Chapter 1). However, where he postulates two stages, parsing and canonicalisation, this model has incorporated canonicalisation into the parsing process by transducing between the levels. Allophonic strings of feature bundles which may be underspecified are input to the PHOPA system. The allophonic redundancy rule component is applied in order to test for consistency and to assign predictable feature specifications which are not contained in the, possibly underspecified, input representation. If the input feature bundles are fully specified, the allophonic redundancy rule component functions only as a consistency checker. The output of the allophonic redundancy rule component consists, therefore, of consistent allophonic feature bundles which are passed to the parser. The application of redundancy rules is described in Section 6.3.2.

The parser interprets the feature-based phonotactic network starting at the initial node transducing between the allophonic and the phonological levels. Every time the parser reaches a final node in the phonotactic network a *possible* syllable or phonological word of the language has been found. Therefore, in order to find more than one phonological word the parsing procedure iterates so that phonological units and syllable boundaries are produced until the input string is empty. The traversal of the phonotactic network is described in Section 6.3.3.

In order to further specify the phonological feature bundle output, the phonological redundancy rule component is consulted before the syllable types and phonological word structures are output. A tree-structured lexicon similar to that proposed by Kay (1983) may then be employed in order to distinguish *actual* words from *possible* words. The movement of the parser through the feature-based phonotactic network may be directed by using the tree-lexicon in parallel. In the cases where the input segment is underspecified, hypotheses are made immediately as to the values of particular features thus excluding paths which will eventually lead to impossible sequences hence increasing the efficiency of the parser.

The basic structure of the PHOPA system is as follows:

INPUT: The input to PHOPA is a string of allophonic feature bundles

OUTPUT: The output from PHOPA is a string of phonological features with syllable types and phonological word boundaries marked

ALGORITHM OUTLINE: With the non-empty input string

1. Apply the allophonic redundancy rules
2. Interpret the feature-based phonotactic network
3. Apply the phonological redundancy rules

COMMENTS: The PHOPA system has not been designed to be incremental, thus each of the steps is applied to the whole input string.

The individual steps are now described in more detail.

6.3.2 Redundancy Rule Application

Redundancy rule application assigns predictable feature specifications to underspecified segments both at the allophonic and the phonological levels. The application is defined on the basis of subsumption and unification and uses the notion of *most specific generalisation* as the condition for rule application. The most specific generalisation is defined here to be the "largest common subsumer" of two feature bundles. It would seem that unification is too powerful a mechanism for this parser. However, this impression arises from the fact that the system currently uses only a binary feature system. As mentioned above the PHOPA model is defined independently of any particular system and it is only the knowledge components which are constructed in terms of a feature classification. Thus, if a complex feature geometry based system was substituted in the knowledge components, the processing domain with the unification mechanism would remain the same.

The basic structure of the redundancy rule application algorithm is as follows.

INPUT: A string of feature bundles

OUTPUT: A string of consistent feature bundles

KNOWLEDGE BASE: The redundancy rule component with rules of the form X $\Rightarrow$ Y where X is the condition for application

ALGORITHM OUTLINE: For each feature bundle in the input string

1. Find the most specific generalisation of the feature bundle and the right hand side of the redundancy rules in order to find the appropriate rule
2. Unify the feature bundle with the right hand side of the redundancy rule

As can be seen here, redundancy rule application checks the consistency of the input feature bundles. If the input to the redundancy rule application is fully-specified, this algorithm acts as a consistency check on the feature bundles and the output is the same as the input. If the input is underspecified, the output is a more fully specified version of the input. This is one of the aspects of the PHOPA model which makes it robust.

6.3.3 Traversal of the Phonotactic Network

Due to the fact that the feature-based phonotactic network is procedurally neutral, depending on the interpretation, it can have two possible procedural aspects; it can be interpreted using one of two strategies: *allophonic-phonological* or *phonological-allophonic*. Thus, a processor which uses the allophonic-phonological strategy performs analysis and a processor which uses the phonological-allophonic strategy performs synthesis. The phonotactic network can therefore be used as a linguistic knowledge base either for a parser or for a generator of possible syllables (or phonological words) or a language.

The strategy used for parsing here is nondeterministic (with backtracking), depth-first and in the allophonic-phonological direction. The processing of the consistent allophonic feature bundle string is undertaken by a finite-state transducer which uses a restricted unification as part of its mapping function. The parallels between the definition of a declarative feature-based phonotactic network and finite state automata were discussed in Chapter 5 (cf. in particular Section 5.3). The extension of the finite automaton interpretation to a two-tape finite state transducer interpretation involves the specification of a δ-mapping between the allophonic feature bundle tape and the phonological feature bundle tape via the category features. The arcs of the feature-based phonotactic network can then be interpreted directly.

The basic structure of the parsing algorithm is as follows:

INPUT: A string of consistent allophonic feature bundles

OUTPUT: A string of phonological feature bundles with syllable or phonological word boundaries marked

KNOWLEDGE BASE: The feature-based phonotactic network with arcs of the form $< n_i, \mathrm{C_a}, \mathrm{C_p}, \mathrm{n_j} >$ defined as in Section 6.2.2

ALGORITHM OUTLINE: The traversal of the phonotactic network is as follows:
Assign the current node $\mathrm{n_i}$ to 0

1. IF the input string is not empty
   THEN for the first feature bundle A in the input string
   2. IF there exists an arc description $<\mathrm{n_i}, \mathrm{C_a}, \mathrm{C_p}, n_j>$
      THEN IF there is a unification U of $\mathrm{C_a}$ and A
         THEN P is the unification of $\mathrm{C_p}$ and $(\mathrm{A}/\mathrm{C_a})$
            Output P
            IF $\mathrm{n_j}$ is the final node *end*
            THEN output a syllable boundary
            assign current node $\mathrm{n_i}$ to $\mathrm{n_j}$ and goto 1
            ELSE goto 2
         ELSE enforce backtracking via $\mathrm{n_i}$
      ELSE IF $\mathrm{n_i}$is the final node *end*
         THEN halt
         ELSE fail
   ELSE halt.

$(\mathrm{A}/\mathrm{C_a})$ refers to the feature bundle containing those features of A which are not contained in $\mathrm{C_a}$. This component also contributes to the robustness aspect of the PHOPA model. Unification, in addition to verifying the well-formedness of the allophonic description, not only discards the allophonic specifications, but also enhances the specification of the input with structural, contextually conditioned phonological information $(\mathrm{C_p})$ from the feature-based phonotactic network.

6.3.4 Example Parses with PHOPA

The following examples were selected to show the work of the phonological parser in two instances: with fully specified input and with underspecified input. The first example (Figure 6.3) for the fully specified input, taken from Carson-Berndsen et al. (1989), is the German word *Postpaket* (Eng: parcel sent by post).

phonetic input matrix

voc	–	+	–	–	–	+	–	+	–
cons	+	–	+	+	+	–	+	–	+
high	–	–	–	–	–	–	–	–	–
back	–	+	–	–	–	+	–	–	–
low	–	–	–	–	–	+	–	–	–
ant	+	–	+	+	+	–	–	–	+
cor	–	–	+	+	–	–	–	–	+
round		+				–		–	
tense		–				+		+	
voice	–	+	–	–	–	+	–	+	–
cont	–		+	–	–		–		–
nas	–	–	–	–	–	–	–	–	–
strid	–		+	–	–		–		–
son	–	+	–	–	–	+	–	+	–
long		–				–		+	
asp	+		–	+	+		+		+

user input

as abbreviations for feature bundles?

Enter Input String: pHOstHpHakHetH

Pause. To continue press any key

symbol output with boundaries marked by $

Searching...

pOstpaket$

no (other) phonemic symbols matching this matrix found

To continue press any key

Figure 6.3: Example parse of the allophonic representation of *Postpaket*

phonetic input matrix

voc	- +
cons	+ -
high	? ?
back	? ?
low	- ?
ant	? ?
cor	? ?
round	?
tense	+
voice	+ +
cont	- ?
nas	+ -
strid	- ?
son	+ +
long	? ?
asp	? ?

phonemic output matrix

user input

...end of data

Pause. To continue press any key

phonetic input matrix

voc	- +
cons	+ -
high	? ?
back	? ?
low	- ?
ant	? ?
cor	? ?
round	?
tense	+
voice	+ +
cont	- ?
nas	+ -
strid	- ?
son	+ +
long	? ?
asp	? ?

phonemic output matrix

voc	- +
cons	+ -
high	? ?
back	? ?
low	- ?
ant	+ ?
cor	? ?
round	?
tense	+
voice	+ +
cont	- ?
nas	+ -
strid	- ?
son	+ +
long	? +
asp	

user input

Output as feature bundles...

To obtain output press any key

symbol output with boundaries marked by $

Searching...

mi$
mE:$
my$
m9$
mu$
mo$
ma:$
ni$
nE:$
ny$
n9$
nu$
no$
na:$

no (other) phonemic symbols matching this matrix found
To continue press any key

Figure 6.4: Example parse of an underspecified two-segment syllable

The second example (Figure 6.4) has been constructed purely for demonstration purposes since this parser has no front-end supplying it with underspecified data. No assumptions have been made here as to which features would, in fact, be recognised by a front-end. The example consists of a two-segment syllable which has been underspecified to the extent that the only specifications which are known are *nasal* for the first segment and *vocalic* for the second segment. It will be shown below how this underspecified input can be enhanced with further feature specifications based on the linguistic generalisations provided in the knowledge components of the parser.

6.4 PROCESSING OF SPEECH VARIANTS

Since no native speaker of a language consistently adheres to a given standard pronunciation in normal conversation, it is an important attribute of any speech recognition system from a robustness point of view to be able to process such nonstandard information. These nonstandard forms will be referred to in this chapter as speech variants. Speech variants are systematic and may arise as a result of a what is traditionally termed a phonological process (e.g. assimilation) or may reflect some personal or dialectal characteristic of the speaker. Variation based on physical or purely physiological factors is excluded. The standard form is taken here to be the phonological representation in the lexicon (written dictionary form in the sense of Grosjean & Gee, 1987) and speech variants are taken to be systematically learned. In the following sections, a classification of speech variants is made which can be explained on a segmental basis. These speech variants are then automatically incorporated into the feature-based phonotactic network and can then be analysed by the PhoPa system. It will, however, become obvious that the processing of speech variants using a segmental model is not adequate, and that for this the event-based model proposed in the following chapter is more appropriate. However, this is the first approximation to the solution of the projection problem.

6.4.1 Classification of Speech Variants

Speech variants occur as a result of traditional phonological processes, for example, elision, epenthesis, or assimilation, or they can arise in line with a regional or dialectal sound change. The word *process* is somewhat problematic here since at the phonetic level it is unclear whether in fact information is deleted by articulatory hiding (cf. Chapter 2) or features are assimilated in certain contexts. It may be the case that such phenomena arise as a result of an overlap of information (cf. Browman & Goldstein, 1989) within a shorter time span which is equivalent to fast speech. The consequences of this line of argument for the event-based parser would be the translation of all allophonic and variant information into overlap and precedence relations. Work is currently being undertaken in this direction.

However, since the phonotactic network of the PHOPA model is segment-based, speech variants are classified according to three abstract processes based on segments: deletion, insertion and substitution. Each of these abstract processes has a corresponding abstract rule type. A *deletion rule* deletes a whole segment and also can modify feature values in neighbouring segments. An *insertion rule* inserts a whole segment and a *substitution rule* is applied to single features in particular segments having the effect of substituting one segment for another. All rule types require a context consisting of directly neighbouring segments. However, this context can in some cases be empty.

Each of the abstract processes can only occur within a particular range of the syllable. On the basis of German data, the following are the "most probable" ranges for the three processes. Deletion and insertion occur only in the rhyme (peak and/or coda) of the syllable and substitution has the whole syllable as its range (i.e. it can occur in the onset, peak or coda). Thus, the application range for each process type restricts the search space required for phonotactic extension. Since the syllable structure is directly represented in the phonotactic network, the search through the network for variations can be restricted to a particular sub-structure (onset, rhyme, etc. - for both nonreduced and reduced syllables). Considering the syllable structure as a hierarchical structure the optimal starting point for the search is defined on the basis of the application range and the search proceeds in a depth-first fashion through the syllable tree. The optimal starting point for an insertion rule, for example, is the peak and its application range is the rhyme of the syllable. If the search is unsuccessful in the peak then the coda is searched.

6.4.2 The NEXT System

The NEXT (Net EXTension) system extends the PHOPA phonological knowledge base (standard phonotactic network) in order that nonstandard forms which arise in fast speech may be processed by a speech recognition system. Additional restrictions are incorporated into the standard phonotactic network in the form of metarules describing particular phenomena. The NEXT system serves is a useful tool for testing the relevance of particular allophonic information in phonological parsing and its theoretical interest lies in its contribution to the field of speech variant learning. An additional phonologically oriented lexicon development tool, WORDGEN (WORDnet GENerator), which involves specialised *word model* construction may be used in connection with the NEXT system in order to compile a verification network which contains all variants of each standard phonological entry in the lexicon. An overview of the NEXT system is provided in Figure 6.5.

The NEXT system has two knowledge components, a set of metarules and

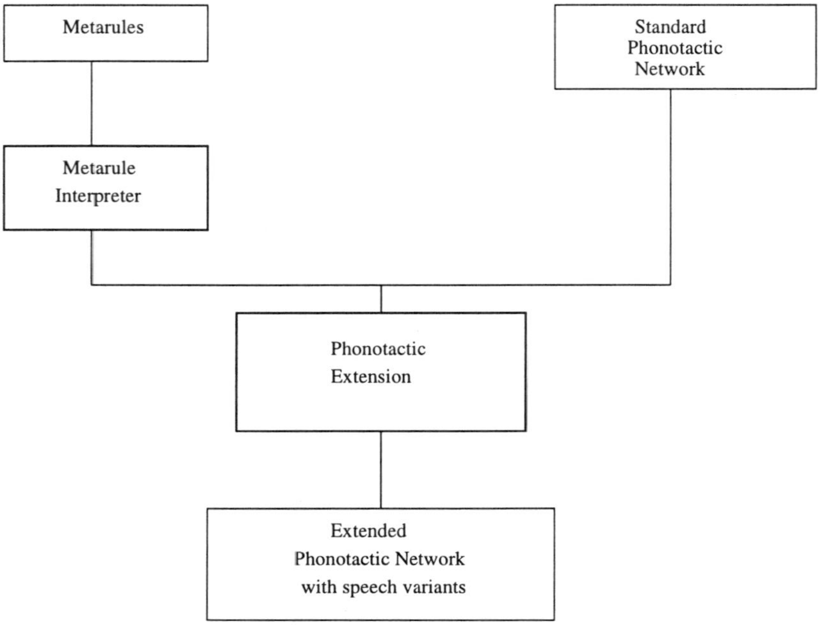

Figure 6.5: Overview of NEXT

the standard phonotactic network. Speech variants are expressed by metarules which describe a particular phenomenon. The metarules have a left hand side and a right hand side, each of which consists of underspecified feature bundles describing well-formed allophonic clusters. The epenthesis of a homorganic voiceless plosive between a nasal and an apical fricative in German, for example, would be described by the metarule in 6.1 which applies to the rhyme.

$$[+nas]\begin{bmatrix} +cont \\ +strid \\ +cor \\ +ant \end{bmatrix} \rightarrow \begin{bmatrix} +nas \\ \alpha ant \\ \beta cor \end{bmatrix}\begin{bmatrix} -cont \\ -strid \\ -son \\ -voice \\ \alpha ant \\ \beta cor \end{bmatrix}\begin{bmatrix} +cont \\ +strid \\ +cor \\ +ant \end{bmatrix} \tag{6.1}$$

This caters for the forms [gamps] for /gams/, [gants] for /gans/ and [gəzaŋks] for /gəzaŋs/. Since declarativity is desirable in knowledge components, the notion of process should be avoided. Metarules should thus be regarded as rules of the form *if form A exists* at the allophonic level, *then form B is possible* at the allophonic level. With such metarules it is possible to describe what would normally be thought of as two processes, namely an epenthesis process and an assimilation process in terms of a single rule.

The extension procedure consists of two stages: interpretation of metarules and phonotactic extension, whereby new transitions are added into the standard phonotactic network. Since in its original form, the metarule cannot be be directly incorporated into the phonotactic network, NEXT first applies a metarule interpreter which produces a graph representation for the metarule. The graph representation corresponds exactly to the syntax of the phonotactic network, that is to say that each transition in the graph consists of two feature bundles: an allophonic specification and a phonological specification.

When the graph representation has been produced, a possible unification is sought for the network and the left hand side of the metarule. This involves a search through the network within the defined range of application. In the case of a substitution rule, for example, a unification would be sought first among the transitions of the onset and then among the transitions of the rhyme. In fact all possible unifications are sought within the application range of the process type, since a metarule may be underspecified and thus may apply in more than one fully-specified context. If the left hand side of the metarule is unifiable with the network, the right hand side of the metarule is inserted into the network at the relevant place by unifying the first and last states of the rule with the relevant states in the network. This phase of the extension procedure results in additional allophonic constraints being added into the phonotactic network.

An additional interpreter which is used in conjunction with the NEXT system is the wordnet generator WORDGEN which compiles specialised word models on the basis of the feature-based phonotactic network. The task of the compiler is to construct for each lexicon entry (i.e. standard phonological form) a corresponding word model which is a verification network containing all variants of the entry. The compiler uses a phonological data-driven processing strategy which transduces between the phonological and allophonic domains. The paths which are consulted in the extended phonotactic network are noted and a verification net for the input word is produced. Word models are therefore subnets of the complete network. Thus, on the basis of a metarule describing a particular phenomenon, the phonological lexicon can learn new variants of a standard wordform. An overview of the WORDGEN system is provided in Figure 6.6.

NEXT and PHOPA together constitute part of the lexical component of a speech processing system. They provide the possibility of analysing some types of nonstandard forms which arise in normal conversation. By adding appropriate metarules, the resulting extended network may be used for parsing nonstandard forms by transducing them into a standard phonological form.

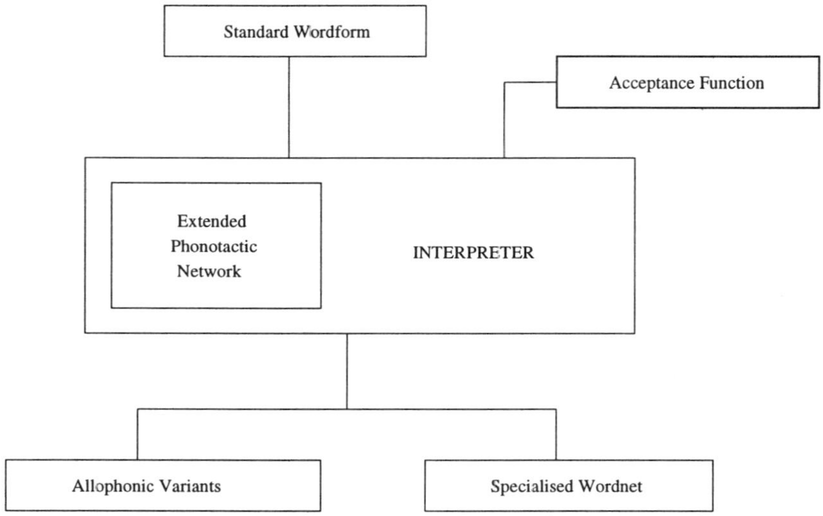

Figure 6.6: Overview of WORDGEN

This chapter presented a constraint-based segmental approach to phonological parsing. In the next chapter, a multi-linear event-based approach is presented which is directed towards the processing of "real" phonetic data. The segmental approach was seen to require a mapping between the allophonic and phonological knowledge domains. The event-based approach goes further by incorporating an explicit notion of temporality, mapping between an absolute time domain T_{abs} and a relative time domain T_{rel} in addition to a mapping between knowledge domains. On the procedural side, the simple depth-first (with backtracking) approach of the segmental parser must be modified in order to cope with the incrementality of real spoken language recognition systems. Furthermore, the straightforward bottom-up strategy employed in the segmental parser is not suited to processing in a multi-component system where other components are able to provide top-down hypotheses which reduce the search space of the parser. These issues are the main concern of the next chapter.

7

CONSTRAINT-BASED PHONOLOGICAL PARSING: AN EVENT-BASED APPROACH

In contrast to Chapter 6, which was primarily concerned with a segmental approach to phonological parsing, this chapter is concerned with a nonsegmental approach to spoken language processing based on the declarative aspects of the event-based approach to phonological description which have been the main theme of this book. The task of this chapter is to present the procedural aspects and in particular an operational semantics for an event-based parser for application in a spoken language recognition system. Furthermore, in order to develop a parser suitable for spoken language recognition a number of issues other than those discussed in the previous chapters must be considered, in particular the notions of incrementality and questions of system architecture in general. Section 7.1 is concerned with processing problems in spoken language recognition which distinguishes it from written language recognition. Section 7.2 introduces the notion of event parsing. In Sections 7.3 and 7.4, the operational semantics for the event-based phonological parser are introduced. In Section 7.5, spoken language recognition system architecture is discussed.

7.1 PROBLEMS IN SPOKEN LANGUAGE PARSING

This section presents some of the problems in the development of a spoken language recognition system which are encountered at all levels of analysis (acoustic to semantic/pragmatic). Usually, language processing refers to the domain of written language. Spoken language processing differs from written language processing in a number of important ways. Written language processing does not usually take account of the temporal domain. For spoken language processing, this is essential, however. Language is processed usually with respect to the category time (T_{cat}) in which categories are assumed to be atemporal. If spatio-temporal relations such as precedence relations are additionally considered then spoken language processing is regarded as a mapping from relative time (T_{rel}) to category time (T_{cat}). However, spoken language processing also makes direct reference to the absolute or signal time (T_{abs}) and therefore spoken language processing for speech recognition or synthesis must be regarded as a mapping from T_{abs} to T_{cat}. The proposal made in this book is that this mapping can most effectively be performed as two mappings: T_{abs} to T_{rel} and T_{rel} to T_{cat}. Written language processing is seldom incremental, although it is possible to construe cases in which this would be desirable (for keyboard input for example), however, the notion of incremental processing outside the spoken language recognition context has been investigated by Wirén (1989, 1990).

Written language (at least above the character recognition level) does not exhibit the same variability as spoken language. This may seem an obvious characteristic of spoken language. However, this characteristic has not been made explicit enough in connection with linguistic analysis and as a result it is often assumed that a processor which works for written language will be sufficient in the spoken language context. Spoken language differs from written language in both its inputs and its outputs. Input to all levels of a spoken language processor may be underspecified, incomplete (with uncertain beginnings or endings) or ill-formed due to noise or to speaker performance. The inputs are not in the form of strings of discrete non-overlapping elements as is usual in written language processing, rather they are lattices containing competing hypotheses and possibly conjunctions in the case of overlapping information. *Competing* in this context does not refer to disjunctions which are usually represented by underspecification but to alternative structures within a temporal interval. Outputs of a spoken language processor are also lattices of competing and conjunctive information. The outputs must also have explicit reference to the signal or utterance time (cf. the indexical interpretation discussed in Gibbon, 1992d).

In the rest of this section, the main emphasis is placed on the issues arising in connection with the temporal dimension of spoken language processing, in particular the treatment of overlaps and gaps in the input lattices. The treatment of such phenomena is required at all levels of processing and has been termed the *lattice-to-chart problem*. This is essentially a level-specific segmentation problem in that a decision has to be made as to whether overlapping information should be regarded as competing hypotheses or as adjacent (or connected) hypotheses. Similarly, in connection with gaps in the input, some provision must be made to allow hypotheses which are separated by a gap to be regarded as connected. The lattice-to-chart mapping is equivalent to making a decision on a particular segmentation (or assuming all possible segmentations) of the input lattice by mapping it to a sequence of non-overlapping hypotheses as illustrated in Figure 7.1.

Recently a number of different solutions (or approximations to solutions) to the lattice-to-chart problem problem have been proposed. Chien et al. (1990) propose a mapping from a word lattice to a linguistic chart based on the hypothesis boundary points (beginning and ending points). The boundary points are sorted into an ordered sequence. On the basis of this ordering the boundary points are assigned to nodes in the linguistic chart. If the boundary point is a

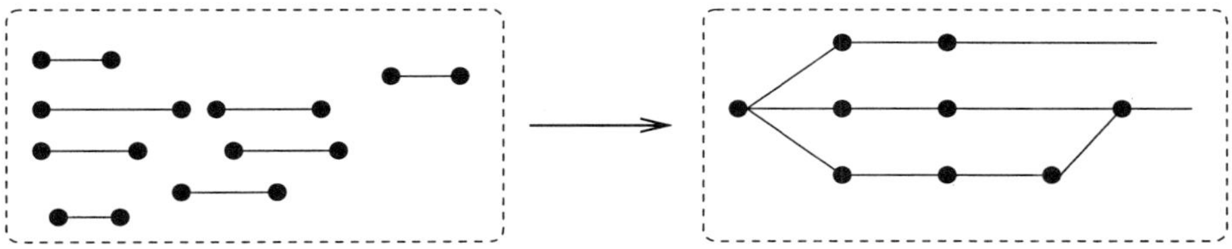

Nonconnected Lattice Connected Chart (or Graph)

Figure 7.1: Depiction of mapping from nonconnected lattice to a connected chart

beginning point and the next boundary point in the ordering is also a beginning point then they are assigned to the same node. Similarly two ending boundary points or an ending boundary point followed by a beginning boundary point are assigned to the same node in the chart. Only in the case of a beginning boundary point followed by an ending boundary point are the boundary points assigned to different chart nodes. Chien et al. (1990) use the notion of *jump edges* for dealing with gaps in the input. However, since this procedure requires that the hypothesis boundary points be sorted before the lattice-to-chart mapping can take place, this procedure is not suitable for incremental processing (unless some sort of window concept is used).

Other solutions to the lattice-to-chart problem which are more suited to incremental processing have been proposed in connection with the ASL speech recogniser at the syntactic, morphological, word prosodic and phonological levels of analysis. In line with Chien et al. (1990), Weber (1992) proposes the use of jump edges for the treatment of gaps in the input lattice. However, he treats overlapping information at the syntactic level by assuming that an overlap of 100 ms refers to two connected (or adjacent) hypotheses whereas an overlap of 150ms is regarded as representing two competing hypotheses. This would seem to be equivalent to the proposal by Posch (1985, cited in Görz 1988:256) where word hypotheses can be regarded as connected if they do not overlap (or are not separated) by more than one segment-length. Clearly there are different measures for each level of analysis. Langer (cf. Gibbon et al, 1992) proposes a solution to the lattice-to-chart problem for the morphological level (the MORPA component) by using chart nodes with "aureoles" which can be interpreted as fuzzy nodes. This approach depends on a parameterised morphological domain time measure. Pampel (forthcoming and Gibbon et al., 1992) suggests parametric temporal granularity measures for the lattice-to-chart mapping in the word prosodic domain which determine whether new chart nodes are generated or whether the boundary points are assigned to the same node. In this way overlapping hypotheses are then either competing or connected. Pampel treats gaps as chart rather than lattice phenomena and adopts an approach using jump edges which are inserted into the chart if no further hypothesis intervenes.

In this book, a somewhat different approach to the lattice-to-chart problem is taken at the phonetic and phonological levels in the syllable processing components of the BELLEx3.1 model (cf. also Gibbon et al. 1992). Rather than performing a segmentation into connected chart nodes, this approach performs a mapping from the temporal annotations or boundary points in T_{abs} to temporal relations between the hypotheses in T_{rel}. Parsing is then carried out entirely in T_{rel} using the relations rather than the temporal annotations. Gaps are described in terms of immediate precedence relations and overlapping hypotheses are described in terms of overlap relations. There are cases where overlapping hypotheses should be regarded as being in a relationship of immediate precedence. This may be calculated according to the percentage of overlap between two events or hypotheses. This issue is currently being investigated further. The next section is concerned with a more detailed description of the *Time Map* solution to the lattice-to-chart problem.

7.2 CONSTRAINT-BASED EVENT PARSING

In Chapter 4, time domains and knowledge domains were introduced in connection with the event concept and Chapter 5 presented constraints on syllable and complex phonetic events. This section is concerned with the role played by time domains, knowledge domains and constraints on event representations in processing spoken language. In the previous section some of the differences between written language processing and spoken language processing were discussed. Spoken language processing was seen to require a mapping from absolute time T_{abs} to category time T_{cat} via relative time T_{rel}. The claim is made here that each analysis level (acoustic to semantic/pragmatic) of a spoken language recogniser corresponding to a knowledge domain requires a mapping from absolute to category time (cf. also Gibbon, 1992d). In other words, it is not the case that only a word level acoustic signal processing component should be concerned with this type of mapping. Furthermore, it is claimed that processing should be done at all levels on the basis of temporal *relations* between events thus avoiding an explicit segmentation into a single symbolic representation. In this way, no contextual information which is important for spoken language recognition is lost or ignored as is the case with approaches which require an explicit segmentation (cf. the arguments provided by Church (1987b); Carson (1988); Kaye (1989); Carson-Berndsen (1991)).

Processing of spoken language in a specific knowledge domain K_j consists of a mapping from absolute time (T_{abs}) to relative time (T_{rel}) at the interface between two knowledge domains K_i and K_j, and a mapping from relative time (T_{rel}) to category time (T_{cat}) which corresponds to the actual parsing of event relations within K_j. Bird & Klein (1990) have provided a basis for the latter mapping ($T_{rel} \rightarrow T_{cat}$) with constraints on possible phonological events but they do not provide a means of relating their events to absolute time. However, in a spoken language recognition system, absolute temporal annotations are provided at the acoustic level on the basis of threshold and confidence values for a particular acoustic event in a speech signal token, and the use of these in calculation of temporal relations for a given signal within an actual speech recognition procedure (the mapping from T_{abs} to T_{rel}) is a nontrivial task which is, in fact, necessary, not an option. In addition to these mappings, each level of analysis (acoustic to semantic/pragmatic) must also be directly related to the absolute time (T_{abs}); the latter is a synchronisation mapping. Figure 7.2 illustrates these stages in spoken language processing in knowledge domains K_i and K_j. T_{abs} corresponds to what was described in Section 7.5 as the horizontal hypothesis search space whereas the vertical hypothesis search space is concerned with both temporal interpretation (a temporal domain mapping: $T_{rel} \rightarrow T_{cat}$) and linguistic knowledge interpretation (a knowledge domain mapping: $K_i \rightarrow K_j$).

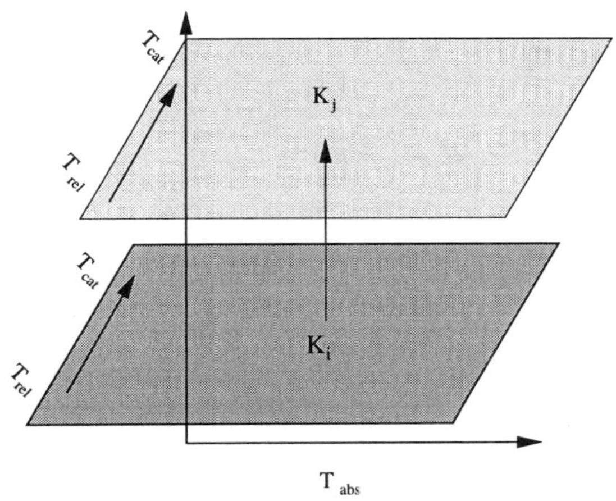

Figure 7.2: Temporal and knowledge domains for spoken language processing

In both domains, there is the possibility of disjunction. Disjunction in the temporal domain gives rise to alternative intervals for a particular hypothesis and disjunction in the knowledge domain give rise to alternative hypotheses for a particular interval. The former relates to the notion of fuzzy points or intervals and is not discussed further in this section (but cf. Section 7.4). The latter is usually expressed by underspecification. The lattices at the interfaces between knowledge domains do not contain either of these disjunction types. In the context of this book, such lattices are termed as *conjunctive competitive lattices* as defined in Chapter 4.

In the next two sections (7.3 and 7.4), an operational semantics for each of the two stages, the $T_{abs} \rightarrow T_{rel}$ mapping and the $T_{rel} \rightarrow T_{cat}$ mapping of constraint-based spoken language processing are presented with respect to an arbitrary knowledge domain K_j. The overall structure is illustrated in Figure 7.3. Examples of how the constraint-based event parser as applied in two knowledge domains are given in Section 7.3.

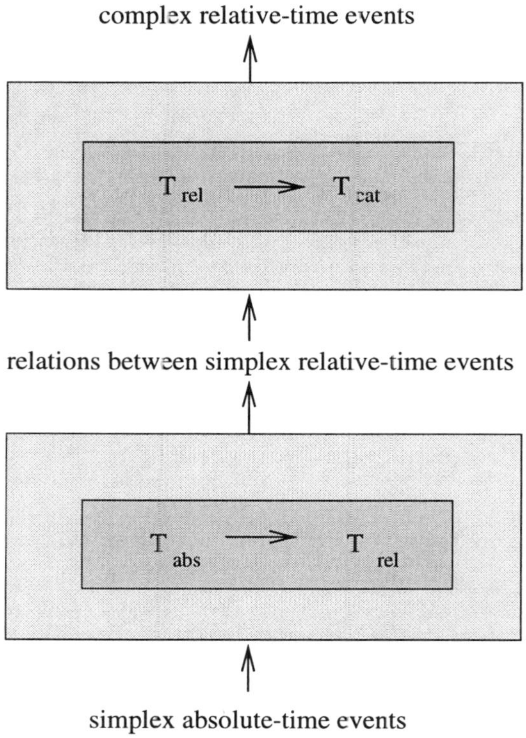

Figure 7.3: Temporal mapping in spoken language processing

The *input* to the $T_{abs} \rightarrow T_{rel}$ mapping is a lattice of simplex absolute-time events as defined in Chapter 4.

The *output* is a set of relations between complex relative-time events which acts as *input* to the $T_{rel} \rightarrow T_{cat}$ mapping. At this stage, knowledge domain-specific constraints are applied in order to restrict the representation by constructing complex events.

The *output* of the $T_{rel} \rightarrow T_{cat}$ mapping is a lattice of complex absolute-time events which is constrained with respect to the knowledge domain K_j. The latter are absolute-time events because the synchronisation mapping has already taken place (they are simplex absolute-time events with respect to knowledge domain K_{j+1}). With respect to incrementality, the output lattice may contain open events representing partial information. This output represents *bottom-up hypotheses* for higher knowledge domains.

Top-down predictions to lower domains (K_{j-1}) form a lattice of open simplex absolute-time events with respect to the knowledge domain K_j. As described in Chapter 4, K_{j+1} and K_{j-1} refer to arbitrary higher and lower knowledge domains in the sense of linguistic granularity. This does not mean that a knowledge domain is only aware of one higher and one lower knowledge domain but rather that bottom-up hypotheses are passed to higher knowledge domains and top-down predictions are passed to lower knowledge domains.

7.3 THE ABSOLUTE TO RELATIVE TIME MAPPING MACHINE

This section is concerned with the mapping from absolute time (T_{abs}) to relative time (T_{rel}) in an arbitrary knowledge domain K_j (in the context of this work the knowledge domains are $K_{phonetic}$ and $K_{phonological}$). The $T_{abs} \rightarrow T_{rel}$ mapping is a procedure which involves two steps:

- temporally relate absolute-time events
- temporally relate relative-time events.

The procedure iterates until all input events events have been temporally related. The basic idea of the $T_{abs} \rightarrow T_{rel}$ mapping is to relate two absolute-time events by inferring from axioms governing temporal relations between absolute-time events. The axioms governing the relations between absolute-time events are based on the signal boundary points (interval end points). When the relation between the two absolute-time events has been inferred, the endpoints are no longer relevant for the relation and they become relative-time events. Further temporal relations holding between all the relative-time events are then inferred, on the basis of the relations constructed so far, from axioms governing the relations between relative-time events. The axioms governing relative-time events are based on the intervals as a whole rather than the signal boundary points of the interval. Absolute-time events are the input to the $T_{abs} \rightarrow T_{rel}$ mapping and the output is a set of relations between relative-time events.

The $T_{abs} \rightarrow T_{rel}$ mapping is first discussed briefly with respect to knowledge structures and inference theory (which have already been presented in Chapter 4) and then with respect to the data structures and the algorithms required for processing.

KNOWLEDGE STRUCTURES: The language-specific knowledge structures required for the $T_{abs} \rightarrow T_{rel}$ mapping were defined in Chapter 4 in terms of simplex events in $K_{phonetic}$ and $K_{phonological}$. The $T_{abs} \rightarrow T_{rel}$ mapping applied in both knowledge domains and is therefore referred to more abstractly as a parameter K_j. A distinction was drawn with respect to temporality between events in T_{abs} and events in T_{rel}. An *absolute-time event* was defined as an ordered pair

$$\alpha\tau\epsilon = << t_s, t_f >, KR > \text{ where } t_s, t_f \in T_{abs}, KR \in K_j$$

and a *relative-time event* was defined as an ordered pair

$$\rho\tau\epsilon = < I, KR > \text{ where } I \in T_{rel}, KR \in K_j$$

Axioms governing absolute-time events in T_{abs} and relative-time events in T_{rel} were proposed.

Absolute-time Axioms (AA): The set of axioms AA = {A1,....,A14} govern the relations between absolute-time events on the basis of their endpoints. The axiom set is provided and extended below. The diagrams illustrate the relations between the absolute-time events; they differ from the diagrams provided in Bird & Klein (1990:50) and van Benthem (1983:70) in requiring endpoint annotations, and in the interpretation of the right-hand margin as "current absolute time". Given two closed absolute-time events $\alpha\tau\epsilon_1 = << t_{s1}, t_{f1} >, \kappa_1 >$ and $\alpha\tau\epsilon_2 = << t_{s2}, t_{f2} >, \kappa_2 >$, the following axioms hold:

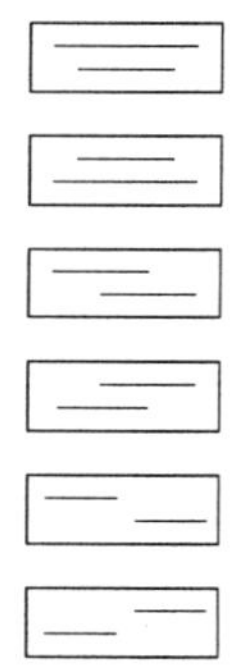

- A1: $t_{s1} \leq t_{s2} \wedge t_{f1} \geq t_{f2} \rightarrow \alpha\tau\epsilon_1 \sqsubseteq_t \alpha\tau\epsilon_2$
- A2: $t_{s1} \geq t_{s2} \wedge t_{f1} \leq t_{f2} \rightarrow \alpha\tau\epsilon_2 \sqsubseteq_t \alpha\tau\epsilon_1$
- A3: $t_{s1} \leq t_{s2} \wedge t_{f1} \geq t_{s2} \rightarrow \alpha\tau\epsilon_1 \circ \alpha\tau\epsilon_2$
- A4: $t_{s1} \geq t_{s2} \wedge t_{f1} \leq t_{s2} \rightarrow \alpha\tau\epsilon_1 \circ \alpha\tau\epsilon_2$
- A5: $t_{f1} < t_{s2} \rightarrow \alpha\tau\epsilon_1 \prec \alpha\tau\epsilon_2$
- A6: $t_{f2} < t_{s1} \rightarrow \alpha\tau\epsilon_2 \prec \alpha\tau\epsilon_1$

The axioms above are relevant for closed absolute-time events. It is also necessary to define axioms for *open events* which are illustrated in the diagrams by the identity of the right-hand point with the current point (i.e. drawing the event line to the right margin). A distinction is made for incremental processing between *right-open events* and *left-open events*. A right-open event is the standard open event as defined in Chapter 4; a left-open event is an event which has finishing time specifications but no starting time specifications. Left-open events provide completion finishing time specifications for right-open events and thus function as closed events. Given two absolute time events $\alpha\tau\epsilon_1 = << t_{s1}, >, \kappa_1 >$ and $\alpha\tau\epsilon_2 = << t_{s2}, t_{f2} >, \kappa_2 >$, where $\alpha\tau\epsilon_1$ is a right-open event, the following axioms hold:

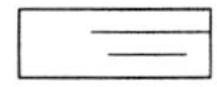

- A7: $\mathrm{t_{s1}} \leq t_{s2} \rightarrow \alpha\tau\epsilon_1 \sqsubseteq_{\mathrm{t}} \alpha\tau\epsilon_2$

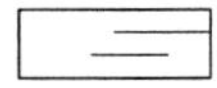

- A8: $\mathrm{t_{s2}} \leq t_{s1} \wedge t_{f2} \geq t_{s1} \rightarrow \alpha\tau\epsilon_1 \,^{\circ}\, \alpha\tau\epsilon_2$

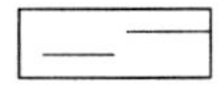

- A9: $\mathrm{t_{s1}} > t_{f2} \rightarrow \alpha\tau\epsilon_2 \prec \alpha\tau\epsilon_1$

Given two absolute time events $\alpha\tau\epsilon_1 = << t_{s1}, t_{f1} >, \kappa_1 >$ and $\alpha\tau\epsilon_2 = << t_{s2}, >, \kappa_2 >$, where $\alpha\tau\epsilon_2$ is a right-open event, the following axioms hold:

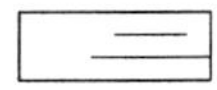

- A10: $\mathrm{t_{s2}} \leq t_{s1} \rightarrow \alpha\tau\epsilon_2 \sqsubseteq_{\mathrm{t}} \alpha\tau\epsilon_1$

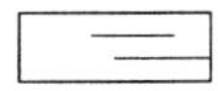

- A11: $\mathrm{t_{s1}} \leq t_{s2} \wedge t_{f1} \geq t_{s2} \rightarrow \alpha\tau\epsilon_1 \,^{\circ}\, \alpha\tau\epsilon_2$

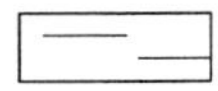

- A12: $\mathrm{t_{s2}} > t_{f1} \rightarrow \alpha\tau\epsilon_1 \prec \alpha\tau\epsilon_2$

Given two right-open absolute time events $\alpha\tau\epsilon_1 =<< t_{s1}, >, \kappa_1 >$ and $\alpha\tau\epsilon_2 =<< t_{s2}, >, \kappa_2 >$, the following axioms hold:

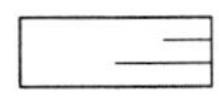

- A13: $\mathrm{t_{s1}} \leq t_{s2} \rightarrow \alpha\tau\epsilon_1 \,^{\circ}\, \alpha\tau\epsilon_2$

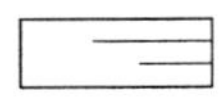

- A14: $\mathrm{t_{s2}} \leq t_{s1} \rightarrow \alpha\tau\epsilon_1 \,^{\circ}\, \alpha\tau\epsilon_2$

Apart from the axioms A1 to A14, there exist *additional* conditions AC which are not axioms, but which describe an overlap relation between closed absolute-time events which may also be regarded as a precedence relation. These are currently formulated as follows:

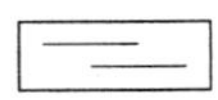

- ac-1: $\mathrm{t_{s1}} < t_{s2} \wedge t_{s2} \leq t_{f1} \wedge t_{f1} < t_{f} \rightarrow \alpha\tau\epsilon_1 \prec \alpha\tau\epsilon_2$

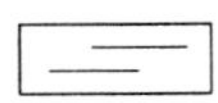

- ac-2: $\mathrm{t_{s2}} < t_{s1} \wedge t_{s1} \leq t_{f2} \wedge t_{f2} < t_{f1} \rightarrow \alpha\tau\epsilon_2 \prec \alpha\tau\epsilon_1$

These additional conditions are equivalent to saying that the axioms governing overlap between closed absolute-time events (A3 and A4) also describe a possible relation of precedence. This corresponds to an extension of the axiom set to modal logic. It is intended that heuristics which take percentage overlap into consideration will be substituted for these additional conditions in a future version of the system (cf. Section 7.4).

Relative-time Axioms (RA): The axiom set RA = {R1,.....,R7} govern the temporal relations between relative-time. For any relative-time events $\rho\tau\epsilon_{\mathrm{i}}$,

- R1: $\rho\tau\epsilon_{\mathrm{i}} \,^{\circ}\, \rho\tau\epsilon_{\mathrm{i}}$
- R2: $\rho\tau\epsilon_{\mathrm{i}} \,^{\circ}\, \rho\tau\epsilon_{\mathrm{j}} \rightarrow \rho\tau\epsilon_{\mathrm{j}} \,^{\circ}\, \rho\tau\epsilon_{\mathrm{i}}$
- R3: $\rho\tau\epsilon_{\mathrm{i}} \prec \rho\tau\epsilon_{\mathrm{j}} \rightarrow \neg\rho\tau\epsilon_{\mathrm{j}} \prec \rho\tau\epsilon_{\mathrm{i}}$

- R4: $\rho\tau\epsilon_i \prec \rho\tau\epsilon_j \rightarrow \neg\rho\tau\epsilon_i \circ \rho\tau\epsilon_j$
- R5: $\rho\tau\epsilon_i \prec \rho\tau\epsilon_j \wedge \rho\tau\epsilon_j \circ \rho\tau\epsilon_k \wedge \rho\tau\epsilon_k \prec \rho\tau\epsilon_l \rightarrow \rho\tau\epsilon_i \prec \rho\tau\epsilon_l$
- R6: $\rho\tau\epsilon_i \sqsubseteq_t \rho\tau\epsilon_j \rightarrow \rho\tau\epsilon_i \circ \rho\tau\epsilon_j$
- R7: $\rho\tau\epsilon_i \prec \rho\tau\epsilon_j \wedge \not\exists \rho\tau\epsilon_k[\rho\tau\epsilon_i \prec \rho\tau\epsilon_k \wedge \rho\tau\epsilon_k \prec \rho\tau\epsilon_j] \rightarrow \rho\tau\epsilon_i \prec^{\circ} \rho\tau\epsilon_j$

INFERENCE STRATEGY: The basic rule of inference for the absolute-time axioms and the relative-time axioms is *Modus Ponens*, that is given a form $P \rightarrow Q$, if P then Q.

The overall structure of the $T_{abs} \rightarrow \mathcal{T}_{rel}$ mapping is illustrated in Figure 7.4.

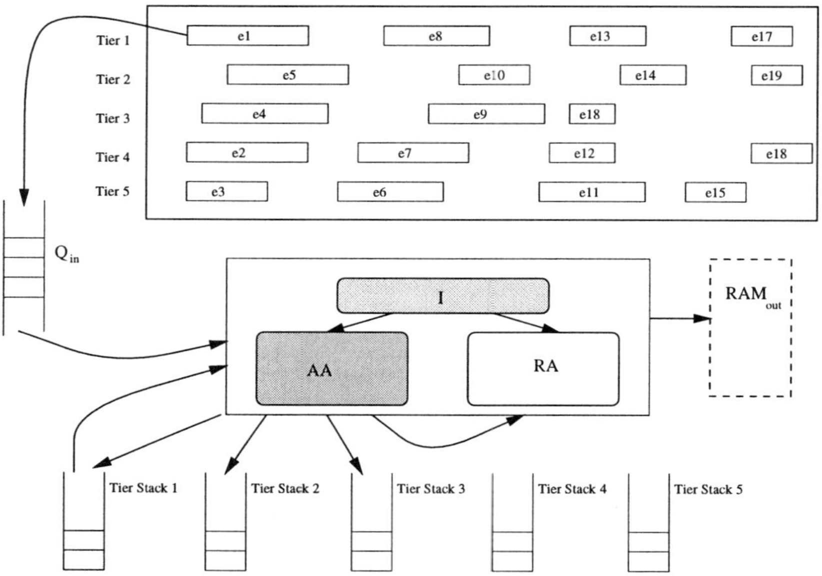

Figure 7.4: The $T_{abs} \rightarrow \mathcal{T}_{rel}$ mapping

DATA STRUCTURES: An abstract machine for the $T_{abs} \rightarrow \mathcal{T}_{rel}$ mapping can be formulated in terms of the following data structures required at each level:

- a read-only queue Q_{in} for the input data (external write)
- a random access memory RAM_{out} for storing relations between relative-time event
- for each tier 1,....i, of the representation a stack *Tier-Stack*$_1$ .... *Tier-Stack*$_i$

The usual PUSH, POP, TOP and EMPTY operations are assumed for stacks and the usual DEQUEUE, ENQUEUE, FRONT and EMPTY operations are assumed for queues. The definitions are taken from Aho et al. (1983:53ff):

- PUSH(x,S): Insert the element x at the top of the stack S. The old top element becomes next-to-top, and so on.
- POP(S): Delete the top element of the stack
- TOP(S): Return the top element of stack S
- EMPTY(S): Return true if S is an empty stack; return false otherwise
- DEQUEUE(Q): deletes the first element of Q
- ENQUEUE(x,Q): inserts element x at the end of queue Q
- FRONT(Q): is a function that returns the first element on the queue Q
- EMPTY(Q): returns true if and only if Q is an empty queue.

For the RAM, the operations PUT and GET are assumed. The RAM is accessed via the unique interval of the first relative-time event in the relation.

- PUT(RAM,I,R):add a relation to RAM at address I
- GET(RAM,I,R):access a relation from RAM at address I

- Elements of Q_{in} are pairs representing absolute-time events
- Elements of RAM_{out} are triples $< e_{\mathrm{i}}, \Re, e_{\mathrm{k}} >$ representing binary temporal relations between events where $\Re \in \{\,^{\circ}, \prec, \prec\,^{\circ}, \sqsubseteq_{\mathrm{t}}\}$. The internal notation for the relations is described below.
- Elements of the $\textit{Tier-Stack}_1, \ldots \textit{Tier-Stack}_{\mathrm{i}}$ are pairs representing absolute-time events.

In addition, the $\mathrm{T}_{\mathrm{abs}} \rightarrow T_{\mathrm{rel}}$ mapping requires:

- a list of (pointers to) axioms AA governing temporal relations between absolute-time events
- a list of (pointers to) axioms RA governing temporal relations between relative-time events
- an inference mechanism I for the axioms

The AA axioms are provided with an internal notation $P^1(t_{ik}, t_{jl}) \rightarrow P^2(\alpha\tau\epsilon_k, \alpha\tau\epsilon_l)$ where $P^1 \in$ {less-or-equal, less, greater-or-equal, greater}, $P^2 \in$ {temporal-inclusion, overlap, precedence} and $\alpha\tau\epsilon_k, \alpha\tau\epsilon_l$ are absolute-time events. The axiom A2, for example, is provided with the following internal representation:

- A2: *greater-or-equal*(t_{s1}, t_{s2}) and *less-or-equal* (t_{f1}, t_{f2}) $\rightarrow$ *temporal inclusion*$(\alpha\tau\epsilon_2, \alpha\tau\epsilon_1)$

The additional conditions ac-1 and ac-2 are provided with an internal notation analogously to the AA axioms.

The RA axioms are provided with an internal notation $P(\rho\tau\epsilon_i, \rho\tau\epsilon_j) \rightarrow (\rho\tau\epsilon_i, \rho\tau\epsilon_j)$ where P $\in$ {temporal-inclusion, overlap, precedence, immediate-precedence} and $\rho\tau\epsilon_i, \rho\tau\epsilon_j$ are relative-time events. The axiom R2, for example, is provided with the following internal representation:

- R2: *overlap*$(\rho\tau\epsilon_i, \rho\tau\epsilon_j) \rightarrow$ *overlap*$(\rho\tau\epsilon_j, \rho\tau\epsilon_i)$

ALGORITHM OUTLINE. The inference mechanism I corresponds to *Modus Ponens*. The rules are interpreted as follows: for axioms of the form *lhs* $\rightarrow$ *rhs*, if the *lhs* is true, then the *rhs* is also true. This is undertaken by the procedure INFER. The following procedures are assumed in the main algorithm below:

```
AXIOMS(AA, RA, ατε1, ατε2):
        For i= 1 to 14
            ABSOLUTE-INFER(Ai, ατε1, ατε2, RHS)
            PUT(RAM_out, I1, RHS)
        For i = 1 to 7
            RELATIVE-INFER(Ri, RAM_out)
```

The procedure ABSOLUTE-INFER corresponds directly to INFER. The procedure RELATIVE-INFER searches the RAM_{out} for the relations required by the current axiom from RA. Further relations inferred by RELATIVE-INFER are added to RAM_{out}.

One further procedure is required before describing the overall algorithm. As a result of incrementality of the complete system, the input absolute-time events may be either closed or open. Appropriate axioms for inferring temporal relations between open events were presented above. However, it is necessary, during processing, to close an open event as soon as the left-open counterpart is encountered in the input. The procedure for doing this is defined as follows:

```
CLOSE(ατε1, ατε2): IF ατε1 =<<, tf1 >, κ1 > and κ1 = κ2
                   THEN let ατε1 =<< ts2, tf1 >, κ1 >
```

The main algorithm for the $T_{abs} \rightarrow T_{rel}$ is as follows:

```
ABSOLUTE-TO-RELATIVE: REPEAT
                      LET ατε1 BE FRONT(Qin)
                      IF EMPTY(Qin)
                      THEN return to REPEAT
                      ELSE IF ατε1= END-MESSAGE
                            THEN terminate
                             ELSE FOR i = 1 to n
                                      LET ατε2 BE POP(Tier-Stacki)
                                      IF CLOSE(ατε1, ατε2)
                                      THEN PUSH(ατε1, Tier − Stacki)
                                      ELSE AXIOMS(AA, RA, ατε1, ατε2)
                      UNTIL FRONT(Qin) = ENDMESSAGE
```

END-MESSAGE is a message sent by a lower component to the effect that there is no more input expected.

As explicitly mentioned in Chapter 4, although the range of the mapping is T_{rel}, this does not mean that the temporal annotations are not visible in the interval I of the relative-time events. Although the temporal annotations are not required for event relation processing, they are needed in order to be able to relate events in higher knowledge domains to T_{abs}.

7.4 THE RELATIVE TO CATEGORY TIME MAPPING MACHINE

This section is concerned with the mapping from relative time (T_{rel}) to category time (T_{cat}) in an arbitrary knowledge domain K_j. This mapping is described less formally than the mapping of the previous section. The reason for this is that the $T_{rel} \rightarrow T_{cat}$ is related to traditional parsing tasks for which there are many well-known algorithms. The approach taken in syllable-based parser of the BELLEx3 system which is described briefly in this section is a variety of an active chart parser. However, it is unconventional in that its uses a network representation of underspecified feature structures as overlap and precedence constraints and in that the interpreter for this network representation is controlled by a finite-state device. An abstract interpretation of the constraint network in terms of a finite automaton was presented in Chapter 5. The basic finite-state automaton has been augmented by additional operations which allow for a breadth-first, predictive parsing strategy. Partial results are stored in a well-formed sub-path table, which differs from a traditional chart in that when it is updated after each input scan or completion, information is emitted to other modules as appropriate and if no longer required, is deleted. This well-formed sub-path table is referred to as an *event-chart* in the rest of this chapter.

The $\mathrm{T_{rel}} \rightarrow T_{cat}$ mapping uses a breadth-first, predictive parsing procedure which involves an iteration of three main steps: EVENT-PREDICT, EVENT-SCAN and EVENT-COMPLETE. The $\mathrm{T_{rel}} \rightarrow T_{cat}$ mapping uses as its declarative knowledge base a network representation of constraints on temporal relations in complex events (cf. Chapter 4). The basic idea is to proceed left-to-right (through the network) and, at each stage in the process, predict top-down the constraints which are applicable (i.e. the next nodes), scan the input for a fulfilment of the predicted constraints provided by the network arcs and complete by allowing the process to proceed to the next node in the network. EVENT-PREDICT uses a strategy which is termed *everywhere-predict* (cf. Section 7.5). This means that in addition to predicting all possible next nodes (standard prediction), the initial nodes are also predicted (initial prediction). This is a hybrid strategy which is neither strictly bottom-up nor strictly top-down. The priority of information varies between bottom-up and top-down. The parser is deliberately kept sufficiently general to allow for more powerful knowledge base interpreters if required.

The $\mathrm{T_{rel}} \rightarrow T_{cat}$ mapping has two sources of input:

- *bottom-up input hypotheses*, a set of relations between simplex events in $\mathrm{T_{rel}}$
- *top-down input predictions*, a set of complex open absolute-time events in $\mathrm{T_{abs}}$.

The top-down predictions are in $\mathrm{T_{abs}}$ due to the fact that they have a direct reference to signal time from the perspective of the higher level.

Analogously, the $\mathrm{T_{rel}} \rightarrow T_{cat}$ mapping has two outputs:

- *bottom-up output hypotheses*, a set of complex absolute-time events in $\mathrm{T_{abs}}$ (i.e the temporal annotations as opposed to the relations are provided)
- *top-down output predictions*, a set of simplex open absolute-time events in $\mathrm{T_{abs}}$.

DATA STRUCTURES: The $\mathrm{T_{rel}} \rightarrow T_{cat}$ mapping can be formulated in terms of the following data structures:

- a random access memory $RAM_{\mathrm{b-u-in}}$ for the bottom-up input lattice
- a read-only queue $Q_{\mathrm{t-d-in}}$ for the top-down input lattice (external write)

- a write-only queue $Q_{\mathrm{b-u-out}}$ for the bottom-up output lattice (external read)
- a write-only queue $Q_{\mathrm{t-d-out}}$ for the top-down output lattice (external read)
- a queue Q_{p} for storing pointers to nodes
- a directed graph G representing a constraint network for the knowledge domain $\mathrm{K_j}$
- a directed graph C representing the dynamic event-chart of well-formed sub-paths through G

An illustration of the structure of the $\mathrm{T_{rel}} \rightarrow T_{\mathrm{cat}}$ mapping is presented in Figure 7.5.

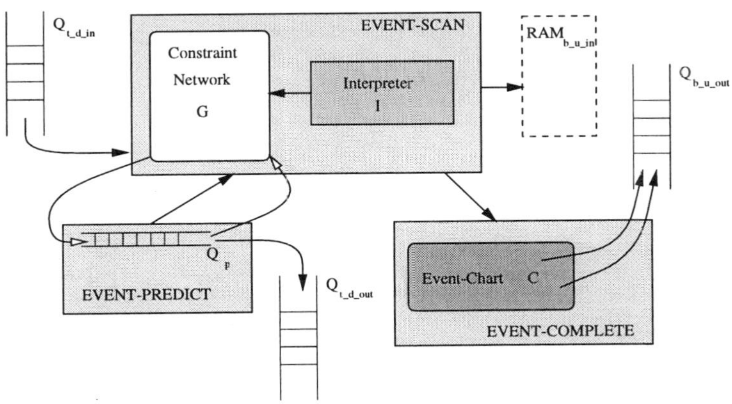

Figure 7.5: The $\mathrm{T_{rel}} \rightarrow T_{\mathrm{cat}}$ mapping

The usual DEQUEUE, ENQUEUE, FRONT and EMPTY operations are assumed for queues as defined in the previous section together with the PUT and GET operations for the random access memory.

- Elements of $RAM_{\mathrm{b-u-in}}$ are triples $< \rho\tau\epsilon_{\mathrm{i}}, \Re, \rho\tau\epsilon_{\mathrm{k}} >$ representing binary temporal relations between events where $\Re \in \{ ^{\circ}, \prec, \prec ^{\circ}, \sqsubseteq_{\mathrm{t}}\}$. The $RAM_{\mathrm{b-uin}}$ corresponds to the RAM_{out} of the $\mathrm{T_{abs}} \rightarrow T_{\mathrm{rel}}$ mapping.
- Elements of $\mathrm{Q_{t-d-in}}$ are pairs representing complex open absolute-time events.
- Elements of $Q_{\mathrm{b-u-out}}$ are pairs representing complex absolute-time events

- Elements of $Q_{\mathrm{t-d-out}}$ are pairs representing simplex absolute time events. Q_{p} contains integers as pointers to nodes of G.
- G is a collection of tuples of the form $< n_{\mathrm{i}}, CON, n_{\mathrm{j}} >$ where $\mathrm{n_i}$ is the first node of the transition (or arc in the terminology of the constraint networks defined in Chapter 5), CON is a list of temporal relations which constrain this graph transition and $\mathrm{n_j}$ is the end node of the transition. The temporal relations have the same format as those in $RAM_{\mathrm{b-u-in}}$.

This is the internal notation for the arc definitions presented in connection with the event-based networks defined in Chapter 5: each arc and each combination of arcs in the constraint network represents a complex event. The arc declarations of the constraint network have the following external representation format: d = $< n_{\mathrm{i}}, C_{\mathrm{E}}, {}^{\circ}, \prec{}^{\circ}, n_{\mathrm{j}} >$ where $\mathrm{n_i}$ is the name of the first node of the arc, $\mathrm{C_E}$ is a set of events which constrain the arc, $^{\circ}$ represents the set of overlap relations between events in $\mathrm{C_E}$, $\prec{}^{\circ}$ represents the set of immediate precedence relations between events in $\mathrm{C_E}$ and $\mathrm{n_j}$ represents the next node; the event-based phonotactic network currently uses only overlap and immediate precedence constraints (cf. Chapter 5).

- C is a collection of tuples of the form $<\mathrm{n_i}, \mathrm{O}, \mathrm{n_j}, < t_{\mathrm{s}}, t_{\mathrm{f}} >>$, corresponding to G except that only the overlap relations (the list O overlap relations in CON) are inserted into the event-chart and that the temporal annotations for the complex event interval are provided in terms of starting and finishing specifications $\mathrm{t_s}$ and $\mathrm{t_f}$ respectively. These tuples are termed *event-items*.

The following event-chart operations are used: INSERT, DELETE and COMPLETE which are defined as follows:

- INSERT(Item,C):insert Item into the event-chart C
- DELETE(Item,C):delete Item from the event-chart C
- COMPLETE(C,Item):if Item has the form $< n_{\mathrm{j}}, O_2, n_{\mathrm{k}}, < t_{\mathrm{s2}}, t_{\mathrm{f2}} >>$ then for every event-item I of the form $< n_{\mathrm{i}}, O_1, n_{\mathrm{j}}, < t_{\mathrm{s1}}, t_{\mathrm{f1}} >>$ already in the event-chart, append the lists $\mathrm{O_1}$ and $\mathrm{O_2}$ to get a list $\mathrm{O_3}$ and insert an event-item $< n_{\mathrm{i}}, O_3, n_{\mathrm{j}}, < t_{\mathrm{s1}}, t_{\mathrm{f2}} >>$ into the event-chart C.

The three steps required for event relation parsing in the $\mathrm{T_{rel}} \rightarrow T_{\mathrm{cat}}$ mapping are now described informally.

EVENT-PREDICT: predicts the next possible constraints on events on the basis of the queue of pointers nodes of the network G. The events which take part in an overlap relationship are enqueued in $Q_{\mathrm{t-d-out}}$ to be passed as top-down predictions to the level of analysis below. In addition predictions are made for all initial node in G, these are not passed to the level below. EVENT-PREDICT differs from the predict of an active chart parser in that it does not predict whole constituents (in this case paths) but just predicts where the next scan can take place.

EVENT-SCAN: For each predicted node n_i of the network G, this procedure searches RAM_{b-u-in} for the constraints specified on each arc which has n_i as its first node. This search is directed by top-down predictions obtained by dequeuing the Q_{t-d-in}; if this queue is empty, then all possible next arcs are tested. This is currently the way in which the local search space is restricted by incorporating top-down predictions. More empirical research is required before exact priorities of local and top-down constraints can be defined.

The constraint network which is structured according to some primary tier (cf. Chapters 3 and 5) is interpreted as a finite automaton by the interpreter I. The mapping from n_i to some n_j is carried out by verifying the constraints specified on the network arc and selecting an appropriate interval for the complex event represented by this arc. The temporal annotations of the intervals of the relative-time events are used to construct an appropriate interval which provides a reference to T_{abs} but this is not discussed further here. Constraints are triples representing temporal relations between events as defined above. The next node n_j of each arc is enqueued in the predictor queue Q_p.

If the scan is successful, the list of overlap constraints O for the arc between m_i and n_j in G are passed to the EVENT-COMPLETE procedure. Important in connection with event-scanning is that positions in the networks are underspecified in terms of event relations, thus indirectly defining a priority between specified and nonspecified event types at those positions. Only the constraints specified on the arcs are relevant for a successful scan, nonspecified constraints may be present in the input but do not affect the success of the scan. This is an important feature of the EVENT-SCAN. The consequences of this approach are discussed further below. The EVENT-SCAN is illustrated in Figure 7.6. The event input to the $T_{abs} \rightarrow T_{rel}$ mapping is given here for purposes of clarity.

EVENT-COMPLETE: this procedure is similar to the active chart parser in that for each successful EVENT-SCAN, it creates a new event-item which is a copy of the network arc (with only constraints on overlap relations). This event-item is inserted into the event-chart using the COMPLETE operation defined above. If COMPLETE is unsuccessful, then INSERT is used. An event-item in the event-chart whose n_i has been designated an initial node of G and whose n_j has been designated a final node of G represents a complex event and is enqueued to $Q_{b-u-out}$ to be passed to the level of analysis above. The events which have led to a successful EVENT-COMPLETE (together with all their relations) are removed from the RAM_{b-u-in}. During EVENT-COMPLETE a redundancy component, similar to that described in chapter 6, optimises the overlap relations before

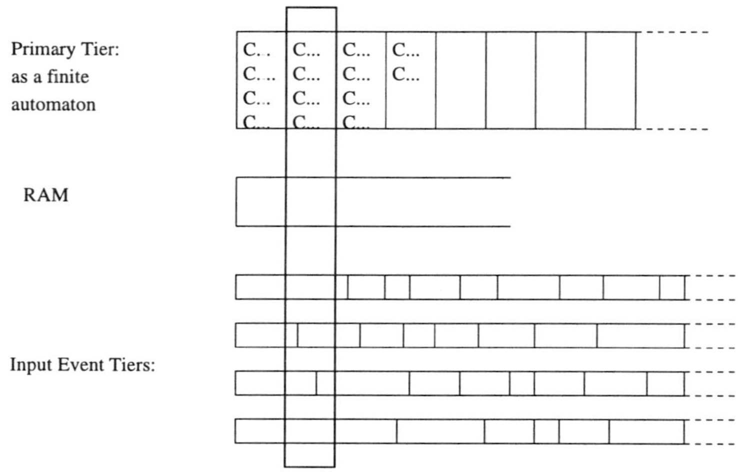

Figure 7.6: An illustration of EVENT-SCAN

inserting them into the event-chart.

These three steps are applied in a cycle until there are no relations left in the $RAM_{\text{b-u-in}}$ and a message to this effect has been received. The parsing process does not fail, it keeps trying for possible complex events (via initial predict).

This section has described the $\text{T}_{\text{rel}} \rightarrow T_{\text{cat}}$ mapping as it has been implemented in the BELLEx3 system. Other types of processor for event relation parsing ($\text{T}_{\text{rel}} \rightarrow T_{\text{cat}}$) would be to have loosely synchronised parallel finite-state machines for each path of the constraint network (distributed processing) or for the input tapes (analogous to the autosegmental machine of Kay (1987)).

In order to be robust, a spoken language processor must allow for *constraint relaxation* (cf. Langer, 1990; Douglas & Dale, 1992). An unrestricted constraint relaxation has been incorporated into the first version of the phonological parser, the COPPA + SILPA solution as described in 8.2. This involves an extension to EVENT-SCAN which allows a minimum of constraints to apply. This type of relaxation is based on the constraint power set: if all the specified constraints for an arc do not apply, the constraint set is reduced from N to N-1. In the N-1 set of constraints there must be at least one overlap relation and one immediate precedence relation. If this constraint set does not apply, it is reduced further until only 2 constraints remain, one of which must be a constraint on overlap and the other a constraint on immediate precedence. This method of constraint relaxation is not restrictive enough (particularly since no explicit ordering is placed on the constraint power set) and is, therefore, not very efficient. A more application-oriented approach to constraint relaxation has been taken in the second version of the phonological parser as described in Section 8.3. However, this is a topic which involves considerable empirical research in order to provide heuristics which define which constraints have a preference for relaxation.

Another concept which has been considered in connection with the $T_{rel} \rightarrow T_{cat}$ mapping is the possibility of enhancing the information provided by the specified constraints (*constraint enhancement*). As mentioned above, it may be the case that the event relations in T_{rel} provide more information than is required by the specified constraints. If this additional information contradicts the specified constraints it should not be allowed to effect the success of the EVENT-SCAN (i.e. should not be considered further). In cases where this information does not contradict the specified constraints it can be used to specify the output further. An example of the first case is a phenomenon which was found during the labelling of the EUROM-0 speech data in the SAM project (ESPRIT 2589 cf. Braun, 1991b): the cluster [szs] in the German word [vɛːʁʊŋszsteːm] as a pronunciation of /vɛːʁʊŋszʏsteːm/ *Währungssystem* (cf. also Carson-Berndsen & Gibbon, 1992). This combination does not correspond to the phonotactics of German because the *fricative* event is dominant throughout the cluster. However, if this is a case of a *hidden vowellike event* in the sense of the Browman & Goldstein *hidden gesture* (1986, 1989, cf. also Chapter 2) and the specified constraints of the phonotactics require only a *vowellike* event, the *fricative* event at this position will be ignored. In the case where the input does not contradict the constraint specifications on a particular arc, information which overlaps with the specifications can be used to further specify the output.

Constraint relaxation and constraint enhancement provide the basis for the processing of speech variants since these often do not correspond to the well-formedness constraints of the language. This offers an alternative to the segment-based classification of speech variants described in Section 6.4 whereby the speech variant phenomena are described declaratively in terms of constraints on overlap and precedence which can be relaxed or enhanced during processing. From the point of view of declarative phonology, this is a more attractive solution to the projection problem in spoken language recognition. Some ongoing research on these issues is discussed in Section 8.4.

7.5 ARCHITECTURE

Speech recognition systems distinguish in general between a *word recognition component* and a *linguistic component*. The linguistic component is responsible for analysis above the word level. Analysis below the word level is considered to be a bottom-up search task. The word recognition component then passes its results to the linguistic component which uses both syntactic and semantic information interactively to interpret the utterance. This type of system can be realised with a relatively simple sequential uni-directional communication architecture. However, speech recognition systems have become more modular; some systems even incorporate linguistic knowledge below the word level. Modularity, in turn, requires investigation of more elaborate architectures and communication both bottom-up and top-down between the modules of the system. There are a number of possible architectures which could be, and indeed have been, employed in speech recognition. These architectures have been discussed in recent literature in connection with spoken language recognition systems (cf. Craig, 1989; Görz, 1988; Sagerer, 1990; Görz, 1992; Görz & Weber, 1992; Pyka, 1991) and it is on these studies that much of this overview of architectures is based.

The first approach is termed the autonomous linear sequential architecture (cf. Figure 7.7), of which the most trivial case is the uni-directional communication mentioned above. The autonomous linear sequential architecture model allows only communication between immediate neighbours in a system.

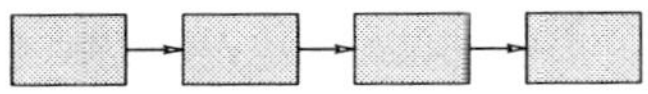

Figure 7.7: Autonomous linear sequential architecture

This approach assumes an ordering between modules of the system and the flow of information is therefore strictly bottom-up. There are two possible search strategies which can be used in connection with such a model; either depth-first (with variants such as best-first) in which case only one solution at a time is passed to the neighbouring module or breadth-first (with variants such as beam or a-best search) in which case all possible solutions are passed to the neighbouring module. This model is too restrictive for speech recognition. A depth-first strategy may result in forcing other modules to work with a solution which later turns out to be incorrect and a breadth-first strategy may result in a combinatorial explosion if all modules find many solutions. For this reason, among others, a cascaded architecture (cf. Figure 7.8) has been suggested which allows a bilateral communication between linearly ordered modules.

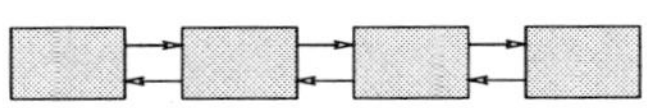

Figure 7.8: Cascaded architecture

This type of architecture allows a module to restrict the search space of its immediate neighbours by providing bottom-up and top-down information. A more interactive version of this type of architecture is the heterarchical architecture model (cf. Figure 7.9) in which communication between models is unrestricted; each module has a bilateral communication with every other module.

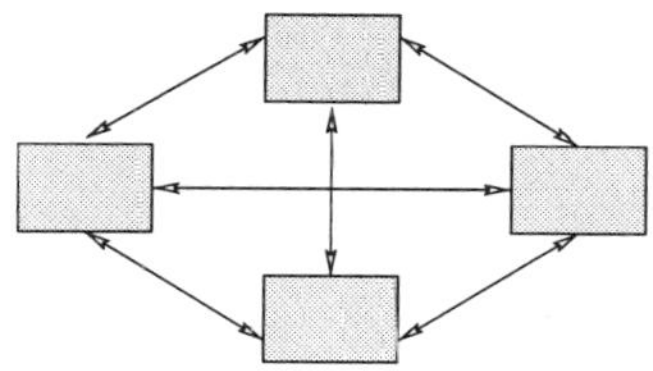

Figure 7.9: Heterarchical architecture

This architecture model incorporates the cognitive notion of functional parallelism described by Marslen-Wilson (1987) in connection with lexical selection using the COHORT model. However, Marslen-Wilson did not assume that the number of modules was unrestricted (n modules ranging from phonetics to pragmatics). The realisation of this type of architecture becomes more difficult the more modules exist and thus it would seem more appropriate to restrict communication to the necessary channels on empirical linguistic and psycholinguistic grounds.

Another architectural model which was originally developed in connection with continuous speech recognition is the HEARSAY-II blackboard architecture (cf. the overview in Erman & Lesser, 1990). The blackboard architecture (cf. Figure 7.10) has arisen as a result of the metaphor that problem solving should be considered in terms of human experts providing solutions from their particular subject area by writing it on a blackboard which can be used by all. The blackboard architecture is now regarded as a general problem solving architecture. It has three main components: a global database (or shared memory) called the *blackboard*, knowledge sources (equivalent to the modules in the discussion above) and a central control component known as the *scheduler*. The knowledge sources are responsible for adding, deleting and altering information in the database. The control component is responsible for managing the information in the blackboard and for centrally controlling the application of different problem solving strategies. The blackboard architecture imposes a sequential hierarchy of abstraction levels on the central database. An excellent overview of blackboard architecture implementations is presented in Craig (1989:14ff).

Craig (1989) has developed another problem solving architecture (*CASSANDRA*) based on the blackboard model but which uses distributed control. This model retains the concepts of knowledge sources and the blackboard. However, the blackboard is divided into a system of autonomous agents called *Level Managers*. Level managers are controlled locally by a local scheduler (cf. Craig, 1989:96f). In addition, level managers have local matchers and

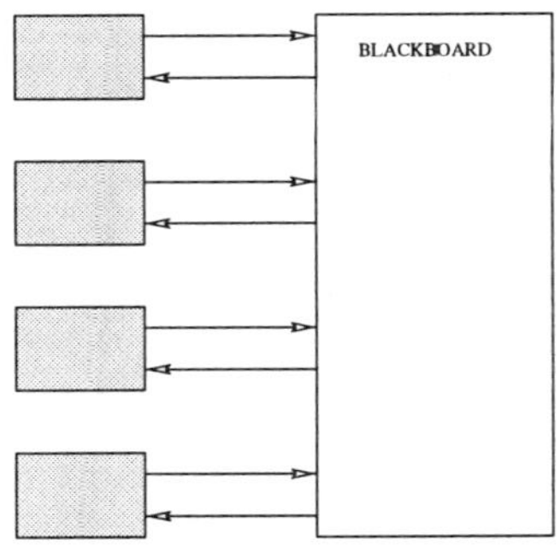

Figure 7.10: Blackboard architecture

action interpreters for the knowledge sources which allow for the possibility of different knowledge representations. Communication is in the form of message passing between level managers which takes place via input or output *ports* along uni-directional *channels* between two level managers. In contrast to the general blackboard, the CASSANDRA level managers can be connected in an arbitrary manner; it assumes no sequential hierarchy.

An architecture model based on cognitive criteria has been developed in the research project *Verbmobil-ASL-Nord* (*Architectures for Speech and Language Systems*, Northern Consortium) which integrates the traditionally separate speech and language components of a system in a homogeneous way. It is this model within which the work described in this chapter was developed. The ASL-Nord architecture model is an incremental interactive blackboard model with decentralised control. It incorporates the functional parallelism of Marslen-Wilson (1987, 1989) and restricts communication between modules of the system to those channels which are cognitively and linguistically plausible. The model must fulfill three cognitive criteria for spoken language architectures: the system should allow for *incremental*, *synchronous*, and *quasi deterministic* spoken language processing (cf. Langer (1991); Pyka (1991); Görz (1992b); von Hahn (1992)). In addition, the system should restrict the amount of working memory required and it should be robust. The architecture is highly modular and assumes bottom-up and top-down interaction via communication interfaces between specified modules. Figure 7.11 shows the structure of the overall system (based on Görz, 1992b:23).

Modularity: This highly modular interactive architecture distinguishes between a horizontal and a vertical search space (cf. Görz, 1992 based on Fodor, 1983) and uses a *radical incremental search strategy* (cf. Görz & Weber, 1991). Incrementality requires modules of the architecture to work with partial or underspecified analyses and therefore allows ambiguities to be resolved interactively. Each module constrains the interpretation.

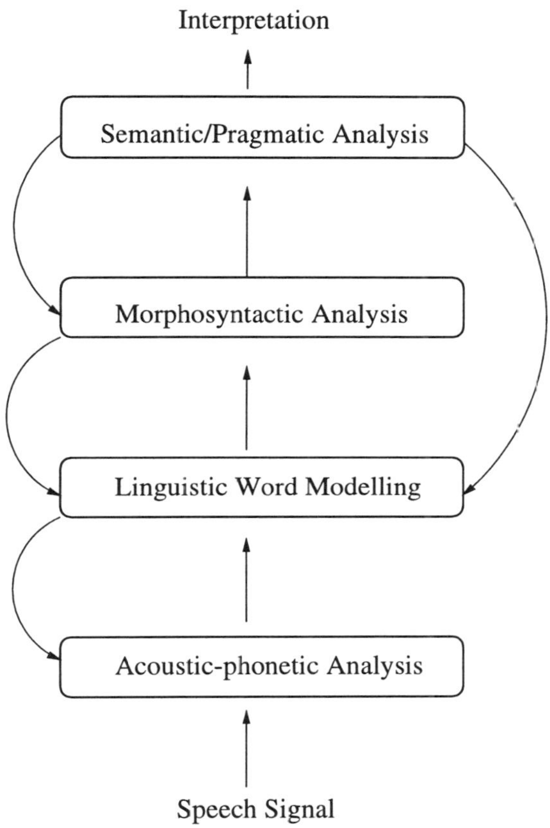

Figure 7.11: Incremental interactive architecture

Incrementality: The radical incremental search strategy proposes that "whenever a partial horizontal search path is expanded try to make it a part of a vertical search path" (Görz, 1992a:7). No module is expected to produce a complete analysis of an utterance before another module can commence processing. Incremental analysis provides the basis for synchronism or on-line processing which is based on the notion that each module of the system is related via some constant to signal time.

Determinism: Determinism refers to the behaviour of the system as a whole, that is to say, the system should only find a single interpretation of the spoken utterance. Determinism is regarded as a long term goal for the system, an approximation, although, individual modules are not deterministic in this sense. For this reason the criterion is referred to as *quasi-determinism*. This architecture module is still under development and there are therefore many aspects of its decentralised control which have yet to be explicitly investigated. The treatment of thresholds and confidence values are but two of these issues.

This chapter has presented the operational semantics for event-based phonological parsing and different possible architectures for spoken language systems have been introduced. In the next chapter two example event-based parsers are presented which have been developed in the context of an architecture model known as BELLEx3.

8

SILPA

In this chapter, two example event-based phonological parsers are presented. Both parsers use the operational semantics described in the previous chapter. They differ in that in the first example in 8.2, the phonetic and the phonological domains have been implemented as two separate processing components COPPA and SILPA. In the second example described in 8.3 the phonetic and phonological domains have been combined into a single processing component - an extended SILPA2 in the context of the BELLEx3.2 system which is introduced below. Section 8.4 presents the development environment for SILPA2 which is used as a workbench for experimentation with different parametrisations of the system.

8.1 THE BELLEX3 ARCHITECTURE

Within the *ASL-Nord* spoken language recognition system, there is a module known as the Linguistic Word Modelling module (cf. Figure 8.1) which has been realised as the BELLEx3.1 system (BELLEx3 Version 1.0). BELLE is the Bielefeld-Engine-for Lattice-to-Lattice-Event-parsing concept (cf. Gibbon et al., 1992). This system is a 3 component module which currently has its own architecture based on the general incremental interactive model. The three components are the morphological parser (MORPA, Langer), the word prosodic parser (PROPA, Pampel) and the event-based phonological parser (COPPA+SILPA, Carson-Berndsen). The system uses a hierarchical background lexicon (Bleiching, 1991, 1992). The PROPA and COPPA+SILPA parsers are designed as multi-tiered processors which account for nonsegmental phenomena. The architecture for the BELLEx3.1 model has been developed according to the linguistic design criteria put forward by Gibbon (1992d). This is discussed further in connection with the phonological parsers in Chapter 8.

It is in principle a heterarchical model in that each component can communicate with all the other components. Each component has two input and two output links corresponding to the spaces for top-down predictions and bottom-up hypotheses. The top-down and bottom-up interactions are shown in Figure 8.1 which is taken from Gibbon et al. (1992).

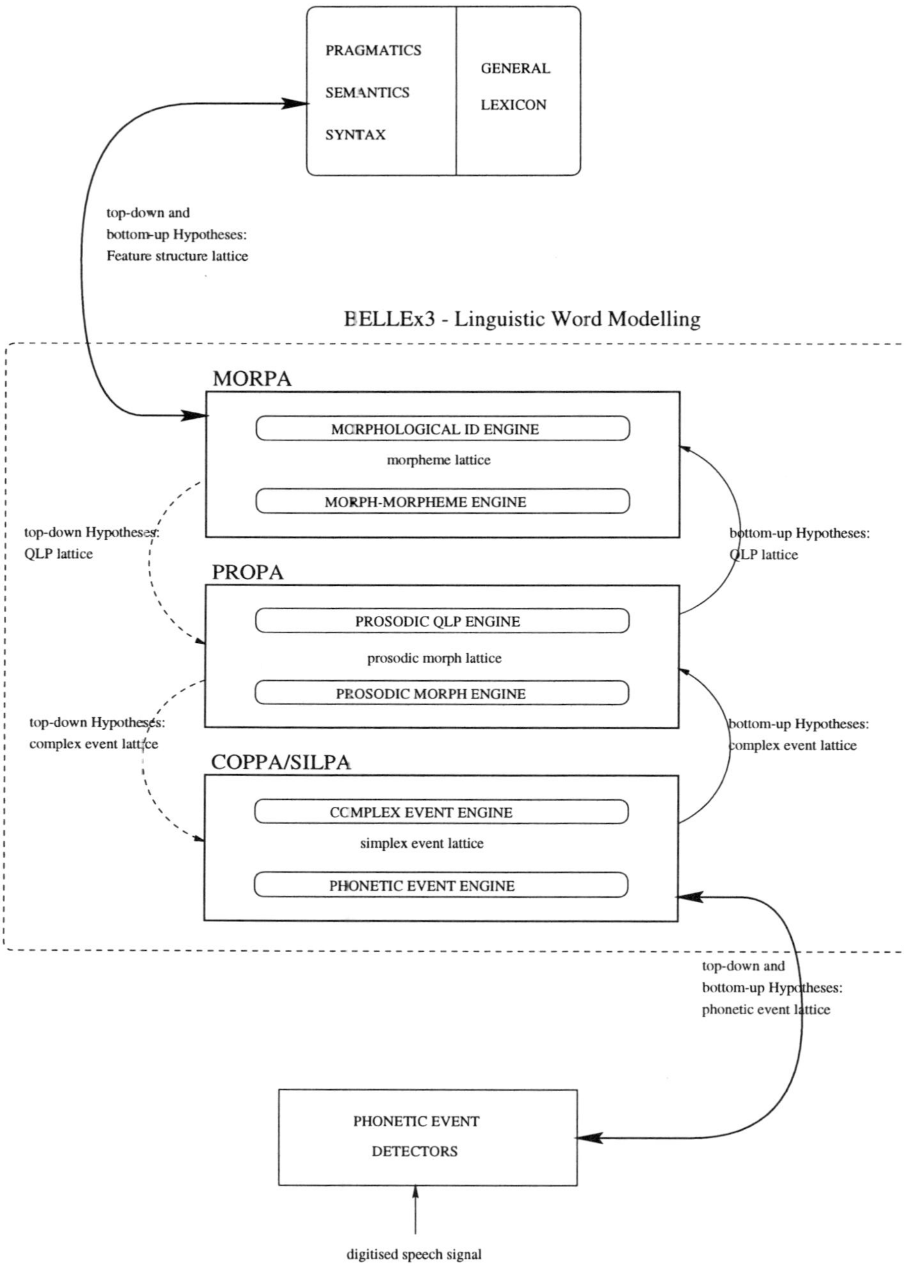

Figure 8.1: BELLEx3.1 architecture

In this way, the components mutually constrain the search space of the other components in the module. Processing is incremental, in line with the general ASL-Nord criterion; partial and underspecified information is made available to the other components. From the point of view that each component makes reference to absolute time, processing can be said to be loosely synchronous. In so far as top-down constraints reduce the local search space, an attempt has been made to maximise the deterministic aspect of processing.

Each component of the BELLEx3.1 model employs a basic iterative algorithm structure of PREDICT, SCAN and COMPLETE. It is a hybrid approach; processing begins with the first *bottom-up* hypotheses (i.e. is initially data-driven) but then incorporates *top-down* predictions. The general search strategy currently employed is breadth-first with an everywhere-predict. Everywhere-predict represents a modification of the PREDICT mechanism of an active chart parser whereby for each standard-predict, in which all the next possible structures are predicted, an initial-predict is carried out in which all possible initial symbols are predicted. The input and output of the components are event-lattices which contain competing information and, in the case of the multi-tiered processing components, also conjunctive information. Each component provides a different solution to the problem of overlapping information which is discussed in the next section. The components have a common interface format which is represented by the tuple:

<Sender, Event, Start-Time, Finish-Time, Confidence-Value>

Event refers to any type of linguistic information which is exchanged between components (e.g. onset events, phoneme events, morph events etc.). Since the components provide partial analyses, the events may be open (cf. Chapter 4) in that the *Finish-Time* is not specified. The Finish-Time is provided whenever the end of the event has been found. Open events are a major feature of this incremental processing model. The treatment of confidence values is a task to be undertaken in a future phase of the project. In the BELLEx3.1 system, confidence values are taken to be equal and therefore there is no hypothesis priority heuristic.

8.2 THE COPPA + SILPA SOLUTION WITHIN THE BELLEX3.1 SYSTEM

This section is concerned with the event-based phonology component of the BELLEx3.1 system. This component consists of two separate parts: the complex-phonetic-event parser (COPPA) for the knowledge domain K_{phonetic} and the event-based phonological parser (SILPA) for the knowledge domain $K_{\text{phonological}}$ (cf. Figure 8.2). Both parsers use the same algorithms and data structures as defined in Sections 7.3 and 7.4. COPPA generates complex-phonetic events (also known as simplex phonological events) from acoustic events using networks describing the well-formedness of complex phonetic events (e.g. plosive, vowellike etc.). SILPA generates syllable and syllable-part events from complex phonetic events using networks describing the phonotactics or well-formedness of syllable events. The output of COPPA is the input to SILPA.

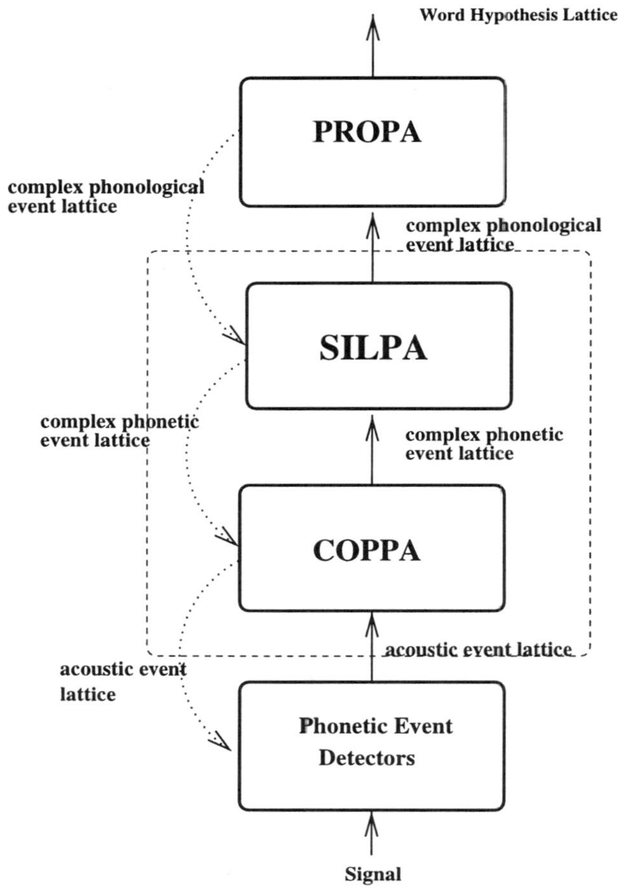

Figure 8.2: The SILPA/COPPA subsection of the BELLEx3.1 architecture

The acoustic event set used in BELLEx3.1 are the input to COPPA. The 13 simplex phonetic events are abbreviated here in terms of their German names, :

sh (voice)
vo (vowellike)
vd (dark-vowel)
vz (central-vowel)
op (occlusion)
fr (frication)
tr (vowel-transition)
st (noise)
vh (light-vowel)
va ([a]-vowel)
sk (syllable-peak)
ol (release)
na (nasality)

It is important to note that this event is logically underspecified in that it is not possible to fully specify all phonological events based on the acoustic event set. This event set must therefore be extended (see Section 8.3).

Events from the phonological event set are the output of COPPA and the input to SILPA. SILPA uses the event-based phonotactics as defined in Chapter 5. For efficient processing purposes, the syllable network given in Appendix C was compiled to a deterministic network and certain loan forms (unusual combinations such as /sf/)were excluded. Syllable and syllable-part events have been passed to the word prosodic parser PROPA. The morph structures which have been constructed on the basis of these events are shown in the PROPA chart. The relevant word structures are constructed in the chart of the morphological parser MORPA. A description of the BELLEx3.1 system is to be found in Gibbon et al. (1992).

8.3 SILPA2 WITHIN THE BELLEX3.2 SYSTEM

The BELLEx3.2 system as a whole differs from the BELLEx3.1 system in a number of important respects. Although BELLEx3.2 still incorporates 3 level-specific processing modules HEAP, SILPA2 and MORPROPA, these modules do not correspond to the modules of BELLEx3.1 as shown in Figure 8.1. The acoustic event detection module HEAP (Kai Hübener, University of Hamburg) has now been incorporated into the BELLE system architecture as a separate processing component. SILPA2 differs from the COPPA + SILPA solution in BELLEx3.1 in that the complex phonetic event parsing and the phonological event parsing have now been implemented within a single processing module. The single parser uses the same algorithms and data structures as defined in Sections 7.3 and 7.4. MORPROPAwhich is a combination of the word prosodic parser PROPA and the morphological parser MORPA of BELLEx3.1 into a single processing component, the morphoprosodic parser (Martina Pampel, University of Bielefeld). The architecture of the BELLEx3.2 system is shown in Figure 8.3.

BELLEx3.2 has been extended to allow the linguistic processing components to apply different strategies when dealing with top-down predictions during incremental processing. BELLEx3.2 has been tested and evaluated on two differing scenarios: railway information and scheduling tasks. The evaluation results are presented in the next chapter in connection with the discussion of evaluation. The phonological and morphoprosodic parsers of BELLEx3.2 are speaker-independent components of a linguistic word recognition component of a speech recogniser which provide linguistic descriptions for the fine tuning of stochastic speech recognition systems.

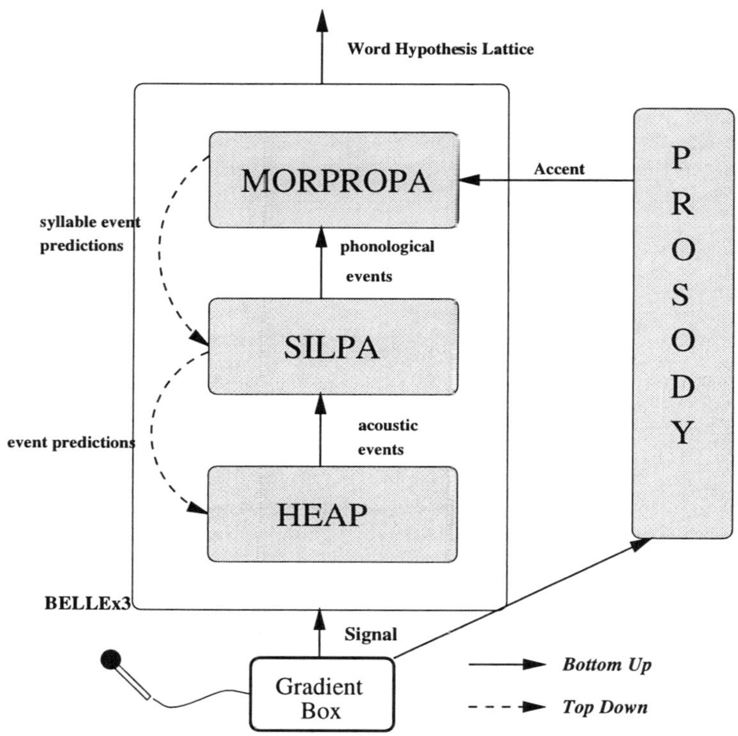

Figure 8.3: Architecture of BELLEx3.2

The acoustic event set which acts as input to SILPA2 was extended to the following 27 events:

pa (pause)	tv (vowel-transition)
vd (dark-vowel)	vr (round-vowel)
vt (low-vowel)	va ([a]-vowel)
vz (central-vowel)	vu (nonround-vowel)
vh (light-vowel)	vg (high-vowel)
vm (mid-vowel)	vo (vowellike)
sk (syllable-peak)	na (nasality)
la (lateral)	sh (voice)
gh (noise)	vo (vowellike)
op (occlusion)	ol (release)
fr (frication)	gl (glottal)
uv (uvular)	ve (velar)
pl (palatal)	po (palato-alveolar)
ap (apical)	lb (labial)

Figure 8.4 shows a screenshot of BELLEx3.2 during processing of the utterance *Ich möchte von München über Nürnberg nach Hamburg fahren.* The SIGNAL window at the bottom left of the screenshot shows how much of the signal has been processed incrementally thus far. The acoustic events which have been detected by HEAP are displayed in the ACOUSTIC EVENTS window. The complex phonetic events which are constructed by SILPA are displayed in the PHONOLOGICAL EVENTS window. The phonotactic network is displayed in the SILPA graphics window in the top right of the screenshot. The darker lines in the network indicate that these arcs are currently active. The status bar (under the title) of the SILPA graphics window highlights the current status in the predict-scan-complete parsing cycle. The output of the word parser is displayed in the MORPROPA graphics window in the top left corner. The word structures are shown in the chart in phonemic form. Here also, the current status of the parser cycle is highlighted in the status bar. The small graphics windows below the SILPA window display the bottom-up and top-down exchange of information between the processing modules. The SILPA and MORPROPA text windows (bottom right) provide status information for the developer.

The BELLEx3.2 system is presented in detail in Althoff et al. (1994, 1995) and the HEAP and SILPA2 components explicitly in Hübener & Carson-Berndsen (1994). The evaluation of the phonological event parser SILPA2 within BELLEx3.2 is discussed in Chapter 9 below.

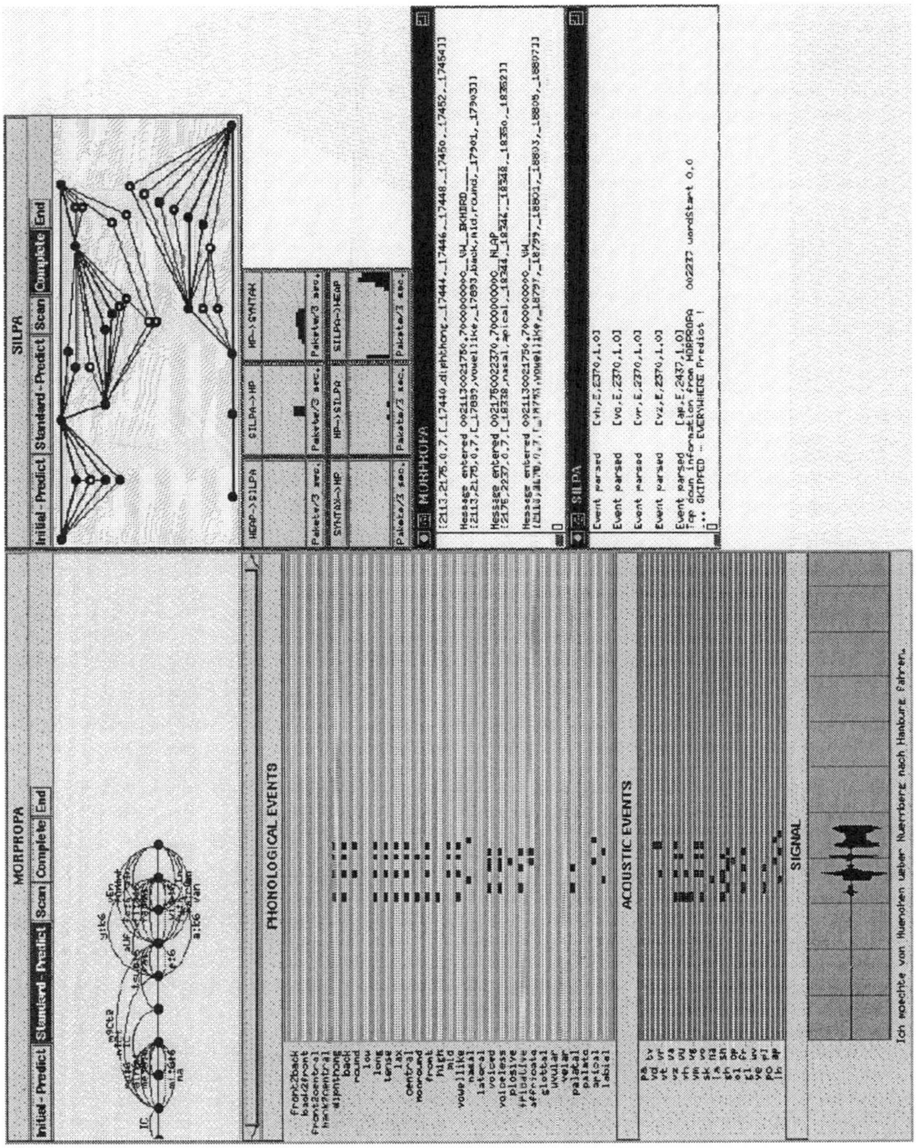

Figure 8.4: Example of BELLEx3.2

8.4 THE PARAMETRISABLE DEVELOPMENT ENVIRONMENT

The phonological event parser SILPA2 in BELLEx3.2 has been provided with an parametrisable development environment (cf. Carson-Berndsen & Drexel, 1996) for examining the extent to which linguistically significant issues such as linguistic competence (structural constraints) and linguistic performance (robustness) can play a role in the spoken language recognition task.

As discussed in Chapter 1, linguistic competence of the native speaker concerns the construction of knowledge bases of linguistic constraints which allow a recogniser for spoken language to distinguish between *actual* structures (i.e. in the lexicon) and *potential* structures (i.e. new words) and to reject ill-formed structures based on information other than the fact that they are not in the lexicon. This is not to claim, however, that a spoken language recognition system should be purely knowledge-based. The long term aim of the work presented here is to provide a linguistic basis for combining knowledge-based and the stochastic approaches to spoken language recognition. In fact, it has already been shown that the explicit incorporation of phonological knowledge can provide useful structural constraints for the fine tuning of stochastic models (cf. Jusek et al. (1994)). With respect to SILPA2 the primary aim has been to examine the role of purely knowledge-based constraints, postponing the incorporation of stochastic constraints until a complete diagnostic evaluation of isolated knowledge parameters has been undertaken.

The development of a model of native speaker linguistic performance presupposes a model of native speaker linguistic competence, since it assumes the *a priori* definition of a norm, from which the native speaker may be allowed to deviate during normal speech. In fact a closer look at spoken language shows that the norm is seldom adhered to; fast speech is a series of words, hesitations, pauses, incomplete utterances, corrections all of which can be filtered by the listener and mapped via a model of native speaker linguistic competence onto an interpretable representation. Quite a lot of research has been done in the area of native speaker performance ranging from work on syntactic normalisation (cf. for example Langer (1990)) to a computational treatment of rule-based speech variation (cf. for example Carson-Berndsen (1990), Kirchhoff (1995a)). It is the issue of native speaker performance which is closely related to the notion of robustness of a system.

SILPA2 can be parametrised according to the following basic classification:

- knowledge base parameters
- performance parameters
- predict strategy parameters

- temporal parameters
- communication parameters
- system parameters

The knowledge base parameters and the performance parameters are the parameters which are relevant for the issues of linguistic competence and linguistic performance. The other parameters in the list will be described briefly before devoting the reset of the discussion to the knowledge base and performance parameters.

The *predict strategy parameters* place constraints on processing within the context of top-down predictions from the word level of processing. A prediction of a word beginning, for example, can be provided by the morphoprosodic parser on the basis of the morphotactic grammar which is then used during phonological processing to direct and constrain the default *everywhere predict* strategy to perform only an initial-predict or a standard-predict, allowing for variations in predictive power which may correspond to memory limitations and other psycholinguistic factors which play a role in shadowing tasks.

The *temporal parameters* are used to define maximum overlap and gap allowances between the events. The upper bound for these parameters corresponds to the maximum length of the phonological word. Those settings for overlap and gap allowances which render the best recognition results provide an indication to the temporal pace and rhythm of the utterance, the temporal aspects of speaker performance at the time of the utterance. However, this issue is part of future research and has not been fully investigated thus far.

Communication parameters and *system parameters* can be termed non-linguistic, that is to say, they do not influence the specific linguistic aspects addressed in this paper. The communication parameters define the method of communication within the complete spoken language recognition system. The system parameters define the format for the output to the user in terms of data files or screen messages, the extent of verbosity of the parser and a confidence value threshold for individual input hypotheses.

The *knowledge base parameters* define the application of constraints on linguistic competence of the model which in contrast to the other parameter types, can be defined independently of the processing system on the basis of strictly linguistic empirical research. Such constraints must offer a solution to the projection problem at the phonetics/phonology interface. Currently, the knowledge base parameters are set to a lexicon containing all (actual) syllables of the corpus and to a complete set of phonotactic constraints defining all well-formed (potential) syllables of German as defined in Chapter 5. However, both these components may be substituted by the relevant syllable lexicon for an English corpus and a system of phonotactic constraints defining all potential syllables of English, for example. The knowledge base parameters thus also responsible for the multilingual aspect of the model.

For a particular scenario, this notion can be taken further. Rather than substituting the phonotactic constraints of one language for phonotactic constraints of another, it is also possible to substitute a system of phonotactic constraints which only covers the scenario corpus. This reduces the functionality of the parser as a computational linguistic model but allows for more efficient processing in a product-oriented system by restricting the knowledge base to more specialised data coverage. The approach presented here allows a ranking of various knowledge bases ranging from the most specific (or purely corpus based) to the most general (or complete language) description.

The *performance parameters* define the application of constraint relaxation and constraint enhancement, modelling the linguistic performance of the native speaker of the language. The performance parameters allow the linguistic competence constraints to be partitioned into the following performance-oriented categories:

1. Obligatory Performance Constraints (OPCs): feature tiers which must be obligatory in the input, i.e. constraints which may never be relaxed; these define the basis for constraint relaxation
2. No Contrary Constraints (NCCs): feature tiers whose features must not contradict a particular specification, i.e. constraints which may be relaxed when no contrary information exists; these define the basis for constraint enhancement
3. Unreliability Performance Constraints (UPCs): feature tiers which can be turned off, i.e. constraints which should be relaxed to the extent that they may be ignored; these define the basis for reducing the importance of unreliable information.

The complete parameter space for these constraints can be empirically verified by iterative tests and evaluation. The optimal parameter settings are defined as those which provide the best recognition results.

Constraint enhancement can be applied using the NCCs which is particularly useful when the input representation is underspecified (noisy). Phonotactic constraints generate a more complete output representation by interpolation as shown in Figure 8.5. In the case where more information is available in the input than is required by the phonotactic OPCs, this information may be used to specify the output further.

Caution is required, however, as unrestricted constraint enhancement (i.e. adding everything into the output which overlaps with the obligatory performance constraints) may overgenerate. For this reason, no contrary constraints

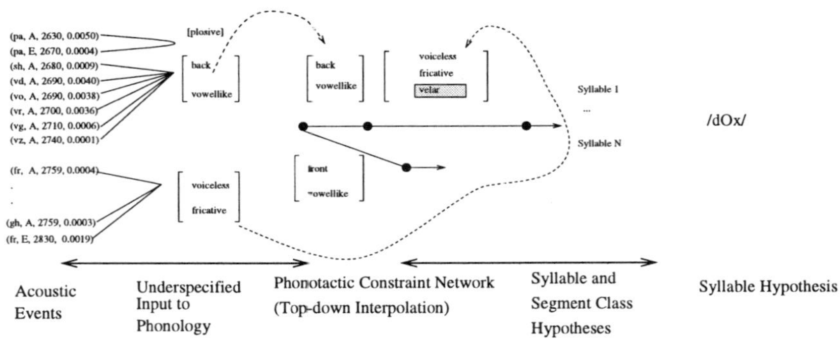

Figure 8.5: Interpolation of Features by Constraint Enhancement

have been introduced which allow only the generation of phonotactically correct forms.

The manner in which linguistic competence constraints are applied can vary and is defined by linguistic performance constraints. The latter are less general, however. In continuous speech human performance is often defective, that is to say, it may contain phenomena such as hesitations, repairs or even syntactically incorrect or incomplete utterances. Continuous input to a spoken language recognition system may be underspecified or noisy and the phenomena found may not correspond to the well-formed structures of the language. It is clear that in order to develop an adequate computational model of phonology, some consideration must be given to robustness issues and the properties of human speech performance.

As mentioned at the end of Section 7.4, a spoken language processor must, in order to be robust, allow for *constraint relaxation* (cf. for example, Langer, 1990). Douglas and Dale (1992) have proposed relaxation levels for constraints in connection with PATR-type rules. They suggest that there should be a distinction between constraints which are necessary and constraints which may be optional. Relaxation level 0 would have no optional constraints, relaxation level 1 would have some small number of optional constraints and so on. Their claim is that the relaxation levels correspond to particular errors (in this case syntactic). A solution similar to this has been investigated in connection with phonological event parsing within the context of the experimental development environment. In the context of a speech recognition system, there are some event types which can be less reliably detected than others. In such cases, it must be possible to relax the phonological constraints and in the optimal case to be able to define the order in which the constraints should be relaxed. In this connection, however, it is not appropriate to describe relaxation levels as corresponding to particular errors in the sense of Douglas and Dale (1992), but rather to attribute constraint relaxation in phonological parsing to native

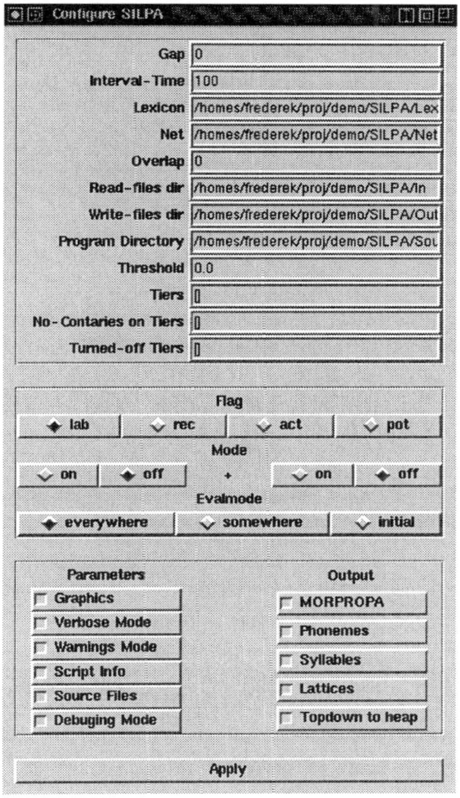

Figure 8.6: SILPA2 Configuration

speaker performance in continuous speech.

As described above, the experimental development environment allows performance parameters to be set which define

1. which constraints may never be relaxed (obligatory performance constraints)
2. which constraints may be relaxed when no contrary information exists (no contrary constraints)
3. which constraints should be relaxed to the extent that they may be ignored (unreliability performance constraints) (i.e. tested in a turned-on or turned-off mode).

Performance parameters define the possibility of specifying various levels of robustness and allow for a distinction in functionality between optimal native speaker performance, where all constraints would be defined as obligatory, and sub-optimal native speaker performance which is the usual case.

The complete parameter space for these constraints can be empirically verified by iterative tests and evaluation and the optimal parameter settings are defined as those which provide the best recognition results. Although the event-based phonological parser is in general speaker-independent, in the case where the input to the system is defined *a priori* as speaker-dependent, optimal constraint relaxation defines speaker-specific performance characteristics for this data. For speaker-independent input, (i.e. many utterances of different speakers), the optimal constraint relaxation defines more general, language-specific performance characteristics such as typical fast speech variants. Other factors such as phonostylistic context would provide useful cues for designing further speaker performance criteria.

Initial experiments for one particular speaker-dependent data set demonstrated that one particular setting of the performance parameters, namely

1. the phonological *manner* tier was defined as obligatory which means that constraint relaxation was allowed over all other feature tiers.
2. the features found on the *place* tier in the input data could not be contradicted by those defined in the phonotactic network, and thus constraint enhancement may only apply when the input values do not contradict the phonotactic specifications - note that this does not effect constraint relaxation.
3. no tiers were defined as unreliable (turned-off).

An improvement in the syllable recognition rate on continuous speech data from an initial 14.5% to 37% was attained. Phoneme recognition which is a side effect of syllable recognition (i.e. derived top-down from the phonemes which occur in the recognised syllables) improved from an initial 49.6% to 72.5%. Diagnostic evaluation and the evaluation of the phonological parser (extended SILPA2 within the BELLEx3.2 system) described in Section 8.3 are the topic of the next chapter.

9

EVALUATION

This chapter is concerned with issues of evaluation. In particular the event-based phonological parser SILPA2 described in Chapter 8 will be evaluated according to criteria which will be defined below. However, before presenting the actual evaluation results, evaluation as a task will be discussed in more general terms. Then a new diagnostic approach to evaluation which has been developed in connection with the BELLEx3.2 system is discussed. This diagnostic evaluation is then applied to SILPA2 and the recognition results are presented in Section 9.6

9.1 EVALUATION IN GENERAL

Evaluation is a measure of the performance of a system. Hirschman & Thompson (1994) distinguish between types of evaluation in speech and natural language processing.

- *Adequacy evaluation* defines the fitness of a system to the task required. This they term evaluation proper.
- *Diagnostic evaluation* is the production of a system performance profile with respect to a taxonomisation of the space of possible inputs.
- *Performance evaluation* measures system performance in one or more specific areas. This type of performance serves as the basis for assessing the progress of a system.

The ESPRIT SAM consortium (cf. Fourcin et al. (1989)) and the EAGLES initiative (Gibbon et al. (1997)) have had the task of proposing standards for the assessment of spoken language recognition systems emphasising the importance of the notions of capability profile as opposed to requirement profile and assessment purpose as opposed to assessment methodology. Consideration should be given also to whether the aim of the system is application-oriented (product-oriented) or research-oriented (development-oriented). The ISO definition is shown in Figure 9.1.

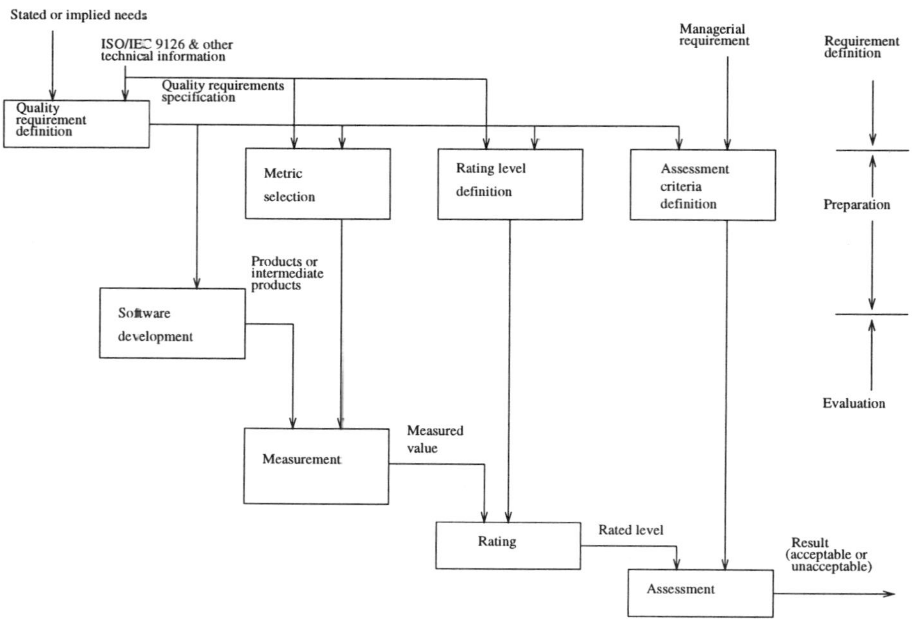

Figure 9.1: ISO Evaluation

Spoken language recognition systems are standardly classified according to the following parameters (cf. Gibbon et al. (1997) Ch. 10):

- Vocabulary size
  - small: $\mid V_S \mid \leq 100$
  - medium: $100 \leq \mid V_M \mid \leq 1000$
  - large: $\mid V_L \mid \geq 5000$
- Speech type
  - isolated words
  - continuous speech
- Speaker dependency
  - speaker dependent
  - speaker independent
- Grammar
  - strict syntax

  - natural language
- Training
  - multiple training
  - embedded training

In the discussion of evaluation measures below, a distinction will be made between *unit measures*, *recognition measures* and *search space measures*. The unit measures are defined with respect to the output of the system and a transcribed reference string:

- *Hit/Correct*: the reference string expected unit_i at position_n and unit_i occurred at position_n in the output.
- *Substitution/Mismatch*: the reference string expected unit_i at position_n but unit_j occurred at position_n in the output.
- *Deletion*: the reference string expected unit_i at position_n but nothing occurred at position_n in the output.
- *Insertion*: the reference string expected the sequence $unit_i\ unit_j$ at commencing at position_n but the sequence $unit_i\ unit_l\ unit_j$ occurred commencing at position_n in the output.

The recognition measures are defined with respect to the unit measures and the number of units N in the reference string:

- $\textit{Error Rate} = \frac{\textit{Deletions+Substitutions}}{N} \times 100$
- $\textit{Correctness} = \frac{\textit{Correct}}{N} \times 100$
- $\textit{Accuracy} = \frac{\textit{Correct}-\textit{Insertions}}{N} \times 100$

The search space measure is defined with respect to the number of units N in the reference string and the total number of hypotheses H_T.

- $\textit{Hypothesis Density} = \frac{N}{H_T}$

A simple evaluation procedure may be defined as follows:

INPUT: a set of hypotheses and a set of references consisting of the units to be evaluated. Both the reference set and the hypothesis set can vary in format.

- **References**:
  - A string of reference units e.g. the man ate the dog (general case)
  - Temporally annotated reference units e.g. $< the, 112, 304 >$ (special case)
- **Hypotheses**:
  - A connected hypothesis graph (general case)
  - Nonconnected hypothesis lattice (special case)

OUTPUT: an evaluation document containing the results with respect to some evaluation measure.

METHOD: compute the similarity between the hypothesis string and the reference string and calculate the required evaluation measure.

A distinction is made here in connection with the references and hypotheses between a general case and a special case. The general case applies to a standard evaluation procedure for a word recognition system. The special case applies to the phonological parsing component SILPA2 described in Chapter 7. Details of the special case will be presented below in connection with a diagnostic evaluation toolkit which has been developed for SILPA2.

In order to compute the similarity between the hypothesis string and the reference string, distance measures are employed. The *Hamming distance*, for example, computes the distance between two strings from left to right without performing an alignment. This is shown in terms of the simple syllable reference string */dɛɒ tɛɒ miːn/* (der Termin 'the appointment') in example 9.1

Example 9.1 *Hamming Distance*

Reference R:	dɛɒ	tɛɒ	miːn
Hypothesis H_1:	dɛɒ	tɛɒ	
Hamming Distance(R,H_1):	1		

Reference R:	dɛɒ	tɛɒ	miːn
Hypothesis H_2:	tɛɒ	miːn	
Hamming Distance(R,H_2):	3		

This distance measure is not suitable for the purposes of the evaluation of a spoken language recognition system since as the example shows, the first hypothesis instance is shown to be closer to the reference string than the second instance, although there is no principled linguistic reason to support this.

A more suitable distance measure which is used in the evaluation of speech recognition systems (cf. Lehning (1994)) is the *edit distance* which requires alignment and weighting and is implemented using dynamic programming. Alignment and weighting are exemplified below.

Alignment is performed on the basis of the following edit operations:

- (a,a) denotes a match
- (a,-) denotes deletion of a in H
- (a,b) denotes substitution of a in R by b in H
- (-,b) denotes insertion of b in H

The examples 9.2 and 9.3 demonstrate alignment using edit operations.

Example 9.2 *Alignment*

INPUT:

Reference R:	dɛɒ	tɛɒ	miːn
Hypothesis H_1:	dɛɒ	tɛɒ	

OPERATIONS:

- correct: (dɛɒ, dɛɒ)
- correct: (tɛɒ, tɛɒ)
- deletion: (miːn, -)

OUTPUT:

Reference R:	dɛɒ	tɛɒ	miːn
Hypothesis H_1:	dɛɒ	tɛɒ	-

Example 9.3 *Alignment*

INPUT:

Reference R:	dɛɒ	tɛɒ	miːn	
Hypothesis H_2:	dɛ	ən	tɛɒ	miːn

OPERATIONS:

- correct: (dɛɒ, dɛɒ)
- insertion: (-, ən)
- correct: (tɛɒ, tɛɒ)
- correct: (miːn, miːn)

OUTPUT:

Reference R:	dɛɒ	-	tɛɒ	miːn
Hypothesis H_2:	dɛɒ	ən	tɛɒ	miːn

After alignment is complete, weighting functions assign values to the errors caused by the edit functions (insertions, deletions and substitutions). These effect the calculation of the recognition measures. Examples of weighting functions are:

- Levenshtein weighting function
  - Insertions = 1, Deletions = 1, Substitutions = 1
  - All errors treated equally
- NIST weighting function
  - Insertions = 3, Deletions = 3, Substitutions = 4
  - Substitutions are penalised more heavily than other errors

Edit distance can be defined as follows (cf. Giegerich & Wheeler (1996)). Note that *s* and *t* represent R and H respectively.

1. The cost of an alignment of two sequences *s* and *t* is the sum of the costs of all the edit operations that lead from *s* to *t*.
2. An optimal alignment of *s* and *t* is an alignment which has minimal cost among all possible alignments.
3. The edit distance of *s* and *t* is the cost of an optimal alignment of *s* and *t* under a cost function *w*. We denote it by $d_w(s,t)$.

The evaluation procedure proposed in this chapter for the evaluation of the SILPA2 component uses a temporal distance measure. This temporal distance measure depends on the notion of a deviation factor which defines by how much hypotheses can deviate temporally from the reference annotations. The examples 9.4 and9.5 demonstrate the application of the deviation factor.

Example 9.4 *Temporal Distance*

INPUT:

R = {<dɛɒ,112,304>,<tɛɒ,304,1254>,<miːn,1254,2679>}
H = {<dɛɒ,104,311>,..,<tɛɒ,345,1288>,..,<miːn,1265,2634>,..}

DEVIATION FACTOR (DV) = 150;

MATCH successful if $| S_H - S_R | \leq DV$ and $| E_H - E_R | \leq DV$

- correct: (dɛɒ, dɛɒ); $8 \leq 150$ and $7 \leq 150$
- correct: (tɛɒ, tɛɒ); $41 \leq 150$ and $34 \leq 150$
- correct: (miːn, miːn); $11 \leq 150$ and $45 \leq 150$

OUTPUT:

Reference R:	dɛɒ	tɛɒ	miːn
Hypothesis H_2:	dɛɒ	tɛɒ	miːn

Example 9.5 *Temporal Distance*

INPUT:
R = {<dɛɒ,112,304>,<tɛɒ,304,1254>, <miːn,1254,2679>}
H = {<dɛɒ,104,311>,..,<miːn,1265,2634>,..}

DEVIATION FACTOR (DV) = 150;

MATCH successful if $| S_H - S_R | \leq DV$ and $| E_H - E_R | \leq DV$

- correct: (dɛɒ, dɛɒ); $8 \leq 150$ and $7 \leq 150$
- deletion: (tɛɒ,-);
- correct: (miːn, miːn); $11 \leq 150$ and $45 \leq 150$

OUTPUT:

Reference R:	dɛɒ	tɛɒ	miːn
Hypothesis H_2:	dɛɒ	***	miːn

The rest of this chapter presents a new diagnostic approach to evaluation which has been developed for the BELLEx3 system described in Section 8.3.

9.2 DIAGNOSTIC EVALUATION

There are a number of motivations for proposing a diagnostic approach to evaluation of the event-based phonological parser SILPA2 within the BELLEx3 system. Firstly, BELLEx3 was designed as a research-oriented system and therefore requires a diagnostic approach to evaluation. Secondly BELLEx3 defines the special case of the evaluation procedure and although the general case of the evaluation procedure can be applied to BELLEx3, this leads to a number of artefacts cf. Section 9.3 which contradict the aims of the system. Furthermore, the components of BELLEX3 must be individually tested for soundness and completeness providing diagnostic information which indicates where improvements can be made in the system.

In the sections below, it is demonstrated how an evaluation procedure based solely on string alignment of connected paths through an output graph using dynamic programming (henceforth *string alignment evaluation*) is not sufficient for evaluating the output of the phonological parsing component SILPA2. Instead a new diagnostic approach to evaluation of individual components within a spoken language recognition system is proposed here which takes soundness and completeness issues into account.

Although this chapter is primarily concerned with the evaluation of the event-based phonological parser SILPA2, the diagnostic evaluation procedure proposed here has been designed to cover also the evaluation of the morpho-prosodic parser (MORPROPA) which forms the other linguistic component of the BELLEx3.2 system. These two components produce three output lattices which are relevant for evaluation: a phoneme lattice, a syllable lattice and a word lattice. The phoneme lattice is regarded as a side-effect of syllable recognition since the phoneme lattice is not a direct product of the phonological parser, but is derived top-down from the phonemes which occur in the recognised syllables. It is important to note, however, that it is not syllable or phoneme hypotheses which are passed from SILPA2 to MORPROPA but rather underspecified subsyllable events. Syllable and phoneme lattices are generated solely for the purposes of evaluation of the SILPA2 component. The output of the MORPROPA component is a word hypothesis lattice. All three lattices differ from the structure of a connected word graph to the extent that the connectedness condition is not a necessary requirement. The output lattices allow the existence of overlapping hypotheses and gaps between hypotheses, a result of the realisation of the original aims of the components, namely the application of a less rigid delayed level-specific segmentation. Although these gaps and overlaps can be interpreted as precedence relations (i.e. connectedness can be forced with the help of an absolute overlap parameter or value relative to which the degree of overlap is defined), the formal criterion of connectedness is not fulfilled from the outset.

9.3 ARTEFACTS OF THE GENERAL CASE

The output of the BELLEX3 components are hypothesis lattices which differ substantially from the connected graph in that they contain overlaps and gaps between hypotheses. It is possible to construct a connected hypothesis graph on the basis of the lattice output by artifically splitting phonological events into phonemic units using temporal statistics and by then mapping the lattice to a chart in the sense of Chien et al. (1990). The constructed chart is then equivalent to a connected hypothesis graph which can be evaluated by a string alignment evaluation procedure. However, this procedure contradicts the aims of BELLEX3 by forcing arbitrary segmentation of events and may also lead to a loss of information.

Example 9.2 shows an example of the overlap of word hypotheses which is typically found in a BELLEX3 output lattice. This must be arbitrarily segmented if a connected word graph is to be constructed.

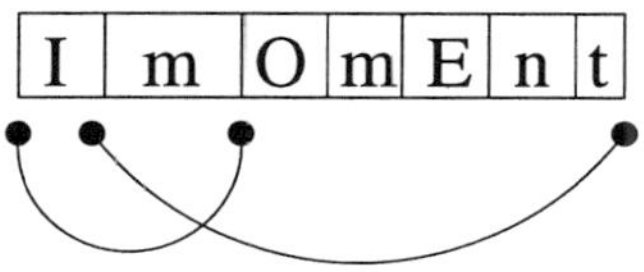

Figure 9.2: Example of overlapping word hypotheses

The phonological event 'm' belongs to two word hypotheses although its temporal duration does not correspond to the realisation of two separate phonemic segments 'mm'. An arbitrary splitting into two phonological events 'm' and 'm' based on temporal statistics is therefore unreliable. This case, in particular, shows the argument against an arbitrary segmentation of temporally overlapping word hypotheses to be convincing.

Figure 9.3 shows a gap between syllable hypotheses within a word hypothesis which is found in a BELLEX3 output lattice.

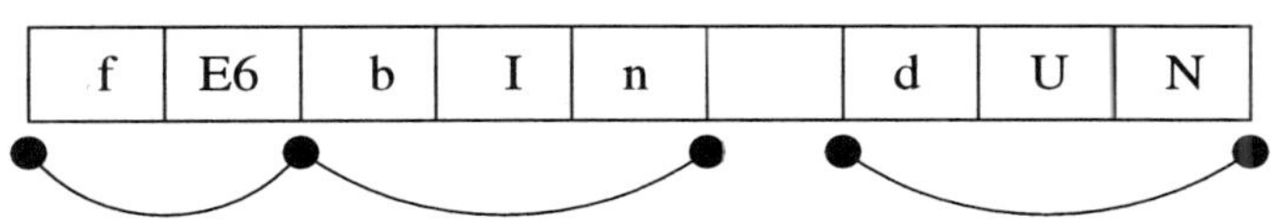

Figure 9.3: Example of gap between syllable hypotheses

This gap must be deleted if a connected syllable graph is to be constructed. However, if any alternative hypothesis occurs during the same interval as the gap this will lead to an incorrect connected graph with a pseudo-insertion of a syllable within the word hypothesis.

Figures 9.4 and9.5 provide more abstract illustrations of overlap and gap. Two possible strategies may be employed in order to be able to apply a string alignment evaluation procedure:

1. either hypothesis **a** can be interpreted as preceding hypothesis **b** (i.e. option (i) in Figure 9.3)

2. or only hypothesis **c** can be chosen (i.e. option (ii) in Figure 9.5).

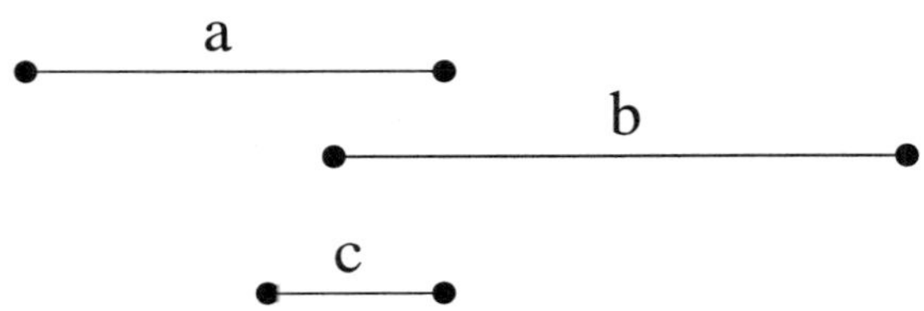

Figure 9.4: Overlap

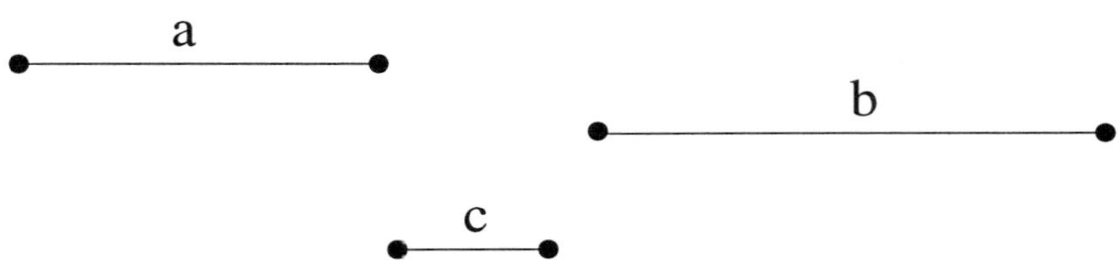

Figure 9.5: Gap

Depending on the overlap or gap factor chosen, certain connected paths may not be constructed resulting in the 'loss' of hypotheses during the lattice-to-chart mapping. Since these hypotheses are not part of a connected path, they will never play a role in the string alignment evaluation. Although a connected graph can be generated from the output lattice, this is counter-intuitive to the fundamental motivation for the late segmentation approach taken by BELLEx3 namely, to explicitly allow exactly the type of structures which the lattice-to-chart mapping discards.

The arguments against a string alignment evaluation for the components of BELLEx3.2 were backed up by the initial evaluation results carried out for the system.The first set of data on which these components were evaluated was that agreed upon by *Verbmobil TP 15: Architecture* for the first INTARC demonstrator in April 1994 (200 utterances of a single speaker from the ASL-*Zugauskunft* (railway information) Scenario). After repeatedly performing evaluation, cor-

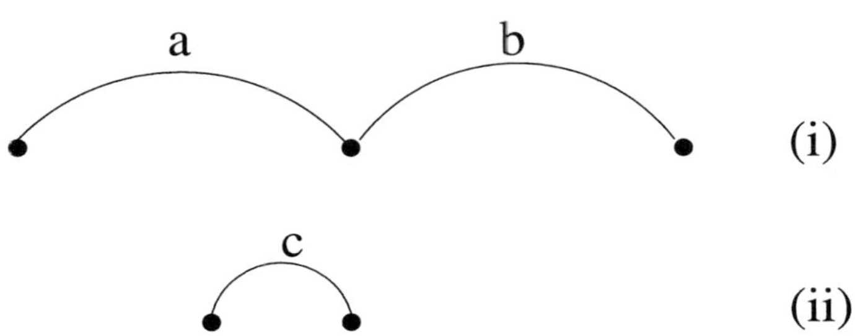

Figure 9.6: Two possible interpretations

recting inconsistencies in the knowledge bases and the parsers, it was discovered that, for the reasons given above, the standard Verbmobil evaluation procedure (Lehning, 1994) produced a recognition rate which was lower than the recognition rate obtained when the output was evaluated manually.

A new diagnostic evaluation procedure has since been developed (cf. Carson-Berndsen & Pampel, 1994) for the SILPA2 and MORPROPA components of the BELLEx3.2 system which takes overlaps and gaps in word lattices explicitly into account. The evaluation procedure can be applied at all linguistic levels, i.e. can be used for evaluation of phoneme output, of syllable output and of word output. Here it is claimed that an evaluation procedure must take signal endpoints into consideration when defining the notion of connectedness. A certain (maximal) amount of overlap between units must be allowed and they may still be considered to be connected. Reference files must therefore contain signal annotations for the utterances, and not merely the utterance as a string. A disadvantage of this approach is that reference files must be generated for all utterances to be evaluated (i.e. regardless of whether the same utterance is spoken by different speakers or several utterances are spoken by the same speaker) which causes difficulties for fast online evaluation of new data. However, since evaluation does not usually take place after each utterance spoken in a demonstration of a system but rather as a collective task on some data set, the temporally annotated reference files can be generated in advance of the actual evaluation.

9.4 STAGES OF DIAGNOSTIC EVALUATION

The primary aim of the work presented here is to go beyond a holistic performance evaluation of components of a spoken language recognition system and enable a diagnostic evaluation of independent parameters. This method of diagnostic evaluation can provide more insights into the role played by linguistically motivated constraints in spoken language recognition than a simple indication of minor changes in recognition rate provided by the holistic approach.

Hirschman & Thompson (1994) distinguish between types of evaluation in speech and natural language processing. Adequacy evaluation defines the fitness of a system to the task required. This they term evaluation proper. Diagnostic evaluation is the production of a system performance profile with respect to a taxonomisation of the space of possible inputs. Performance evaluation measures system performance in one or more specific areas. This type of performance serves as the basis for assessing the progress of a system. This chapter is concerned with diagnostic evaluation in the sense described above, with the addition that diagnostic evaluation is considered to be a more general term which covers both adequacy evaluation and performance evaluation. Diagnostic evaluation is defined as consisting of two evaluation stages: *logical evaluation* and *empirical evaluation*. Logical evaluation is undertaken with respect to a data model. The data model defines one possible input space for the system, namely the optimal input data, and is generated top-down on the basis of labelled speech files. Optimal input data is the input a system would hope for in the ideal case. This concept is relevant only for levels of processing which explicity use structured linguistic knowledge since only here is it possible to define what optimal input would be. At the level of sentence syntax, for example, it is possible to define what the optimal input for a sentence parser would be; a single utterance which is grammatically correct. This will not resemble real input in a spoken language recognition system but evaluating the sentence syntax level with optimal input is equivalent to verifying that all components which participate in the logical evaluation are internally consistent. Linguistic components of speech recognition systems are often criticised for assuming a near optimal input which leads to problems when the parser is coupled with an acoustic component. However, if optimal input is used for evaluation in order to develop a consistent linguistic component which has been designed also to deal with suboptimal input, then an evaluation with real data can be carried out without internal inconsistencies leading to failure. As mentioned above, logical evaluation can be applied at all linguistic levels. In particular, the rest of this chapter is concerned with the evaluation of the phonological parsing component (i.e phoneme and syllable and recognition).

In some ways logical evaluation is similar to testing a stochastic model with training data rather than with test data; such a procedure would not allow for participation in a competitive evaluation of the performance of several systems, but it does indicate the performance levels which can hoped to be attained on real data after tuning has taken place. Logical evaluation of linguistic word recognition components differs from the above in that a recognition rate of 100% can be achieved if the all components which participate in the evaluation are sound and complete. The stochastic model relies on a certain statistical generalisation which makes a 100% recognition rate on training data more difficult.

Empirical evaluation is defined here as evaluation on real input data. It is the recognition rate achieved by empirical evaluation which can be compared with current evaluation results in the area of speech and natural language processing.

9.4.1 Logical Evaluation using a Data Model

Logical evaluation of a component involves a test for soundness and completeness with respect to a data model. In order to perform a logical evaluation the following steps are necessary:

- *Task*: Test all the entries in the lexicon with the grammar of the component.
  Consequence: If the grammar does not permit analysis of all lexicon entries, then either a correction of the lexicon or a revision of the grammar is required.

- *Task*: Generate optimal input data for the component using either automatically phonemically labelled data or manually corrected label files and test the component.
  Consequence: The component must at least be able to analyse what it considers to be optimal data. Otherwise the processing of the component is incorrect according to the optimal data model.

- *Task*: Generate automatically the reference files for the evaluation software using either automatically phonemically labelled data or manually corrected label files. This must correspond to the format chosen in connection with the generation of optimal input.
  Consequence: These files serve as the basis for the evaluation and if inconsistencies are found then evaluation will not be correct.

- *Task*: Test the evaluation software for inconsistencies.
  Consequence: If the evaluation procedure is inconsistent, then a new procedure must be drafted.

- *Task*: Visualise the output lattices of the component.
  Consequence: The user can see the extent of overlap and gaps in the output lattice of the component.

- *Task*: Visualise the output of the evaluation software.
  Consequence: The user can compare this visualisation with that of the output lattices to see whether any information has been lost.

In addition, in order to avoid inconsistencies, the lexicon for the component should also be generated automatically. However, this is a knowledge acquisition task rather than a step in the logical evaluation procedure. As will be seen below, these two tasks are interrelated and important for a successful diagnostic evaluation.

As was mentioned above, the assumption is made here that in linguistic processing, a recognition rate of 100% can be achieved by the logical evaluation. Only a component which achieves this rate is sound and complete for *this* data model. Logical evaluation may therefore be an iterative procedure and it is not until a recognition rate of 100% has been achieved, that an empirical evaluation should be carried out.

After an iterative logical evaluation had been performed for the syllable recognition module on the scenario defined above (200 utterances of a single speaker), a logical recognition rate of 98.9% was achieved using the standard string alignment evaluation software. The error rate of 1.1% is due to the fact that no phonologically and morphologically relevant temporal statistics were calculated for this scenario and therefore long segments were not divided into two separate segments and therefore two overlapping syllables were generated (cf. 9.2). However, a visualisation tool developed explicitly for the purposes of diagnostic evaluation (see Section 9.5.6) indicated that both syllables had been found but that they did not stand in a connectedness relationship. Since this type of phenomenon is likely to occur even more frequently in real data, it was decided to develop an evaluation procedure which caters for the needs of the linguistic components of BELLEx3.2. This evaluation procedure is referred to as the BELLE evaluation in the rest of this chapter.

The BELLE evaluation procedure has been implemented and is described in Section 9.5.5 in detail below. It is based on the notion that a reference file which only defines the connectedness relationship between units is not sufficient for evaluation purposes since signal time also plays a role. The recognised units must also correspond to a subsection of the signal. Temporal annotations for the units of reference files can be defined top-down from label files. Since it is unlikely that a recogniser of these units will recognise precisely these units at precisely these signal time points, a deviation parameter must be defined which specifies by how much the recognised unit may differ from the temporal annotations in the reference file.

The deviation factor varies according to the size of the unit. A syllabic unit, for example, may be permitted to deviate in its endpoints up to 150 ms from the temporal annotations. Phonemic units, on the other hand, would only be allowed to deviate by say less than 30 ms. However, since the correct deviation factors for the respective levels depends on empirical testing, the evaluation software must be parametrised in order that all values can be tested. The deviation parameter which produces the best results is then the most suitable for this unit.

The BELLE evaluation software produced a logical recognition rate of 100% for the phonological parser. Since it was known in advance that only one-segment overlaps occurred, it was possible to set the deviation parameter immediately to 30ms. This will always be the upper bound for the deviation parameter with logical evaluation using a data model. However, as will be seen in the next section, the setting of the deviation parameter plays an important role in empirical evaluation.

9.4.2 Empirical Evaluation and Linguistic Word Recognition

Empirical evaluation also involves most of the steps defined under logical evaluation although the majority of the steps will have been taken in connection with logical evaluation and thus will not have to be repeated. Empirical evaluation differs from logical evaluation to the extent that evaluation of the component is undertaken with real data rather than with optimal data which is generated top-down from labelled data. In addition to the evaluation, it is desirable to have a visualisation of both output and evaluation results. However, it is obvious that empirical evaluation is only meaningful when a complete logical evaluation using a data model with manually corrected labelled data has been performed and a recognition rate approximating 100% has been achieved.

In this case also, the first set of data which was evaluated was that agreed upon for the first INTARC demonstrator in April 1994 (200 utterances of a single speaker taken from the ASL-*Zugauskunft* Scenario). The phonological parser output was obtained with the setting of the performance parameters for SILPA2 as defined in Section 8.3 of the last chapter, namely

1. the phonological *manner* tier was defined as obligatory which means that constraint relaxation was allowed over all other feature tiers.
2. the features found on the *place* tier in the input data could not be contradicted by those defined in the phonotactic network, and thus constraint enhancement may only apply when the input values do not contradict the phonotactic specifications - note that this does not effect constraint relaxation.
3. no tiers were defined as unreliable (turned-off).

An empirical evaluation produced a phoneme recognition rate of 72.2% and a syllable recognition rate of 37%. As mentioned above, the phoneme recognition rate is a side-effect of syllable recognition since the phonemic units are calculated top-down from the recognised syllables. The further empirical evaluation of the morphoprosodic component MORPROPA with this data produced a word recognition rate of 50%.

Before further evaluation results for SILPA2 are presented, a front-end for diagnostic evaluation of SILPA2 is presented which has been developed by the current author at the University of Bielefeld called BEETLE (Bielefeld Extended Evaluation Toolkit for Lattices of Events) [1].

[1]The main implementation of BEETLE has been undertaken by Julie Carson-Berndsen and Frederek Althoff at the University of Bielefeld. Thanks is due also to Guido Drexel, Katrin Kirchhoff, Martina Pampel, Christoph Schillo and Markus Vogt for implementation of individual tools.

9.5 BEETLE

BEETLE is a toolkit for diagnostic evaluation which allows both logical and empirical evaluation to be done automatically given phonemically labelled data[2] in a predefined format, and a phonological parsing component.

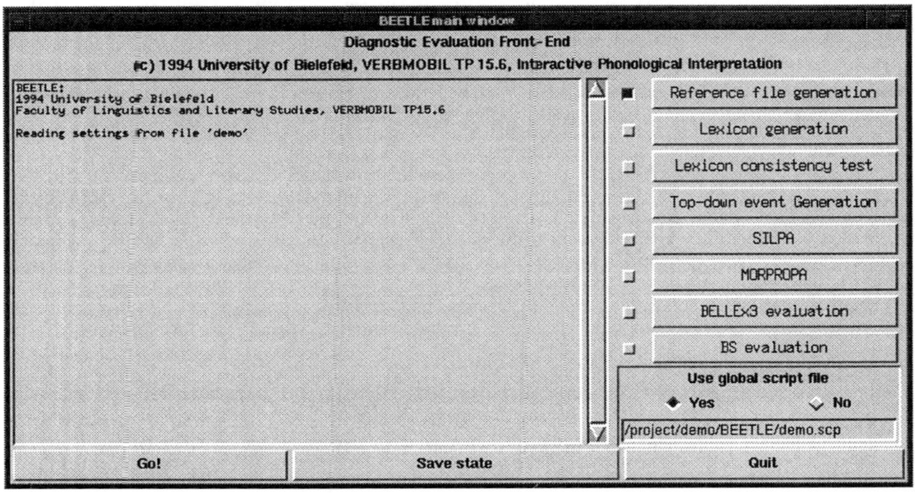

Figure 9.7: BEETLE User Interface

Figure 9.7 shows the main BEETLE window with the possible stages of diagnostic evaluation. In addition to the steps defined in connection with logical evaluation in Section 9.4.1, an additional step has been incorporated which concerns the acquisition of linguistic knowledge. The lexicon for the phonological parser which allows a distinction to be made between *actual (or lexicalised in the corpus)* and *potential* syllables is generated automatically on the basis of the phonemically labelled speech data.

In the following subsections, the individual stages of diagnostic evaluation are described in detail.

[2]The term phonemically labelled data as used in this report refers to either automatically phonemically labelled data or to manually corrected labelled data

9.5.1 Reference File Generation

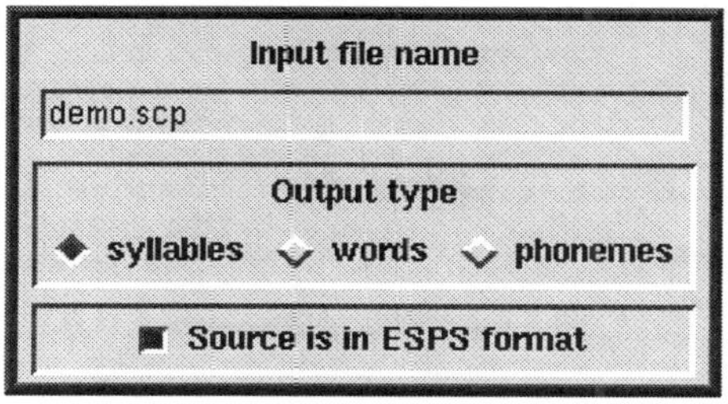

Figure 9.8: Reference File Generation

GOAL: generation of reference files for the output unit (i.e. in the case of SILPA2 these units are phonemes or syllables).

INPUT:

- the unit
- a script file listing the phonemic label files in either the following format:
  Start-Time End-Time Label Confidence-Value
  or in ESPS label file format.

OUTPUT:

1. the general case reference file (for string alignment evaluation)
2. the special case reference file (for the BELLE evaluation).

METHOD: The reference file generation works in the following way:

1. the phonemic label files are converted into an internal format.
2. all possible substrings of the complete utterance are calculated and are presented to the user who marks the required structures using the cursor and return keys.

It would be possible to insert a parser which generates only the relevant strings for the unit (e.g. syllables) but the tool has been kept more general in order that other substructures such as demisyllables or bigrammes can also be marked. An example of a section of such a file is provided in Figure 9.9 where only the relevant syllables are marked. When this marked file is stored by the user, it is converted into the label format defined above providing, in this case, a syllable label file. On the basis of this second label file, the two reference files are generated.

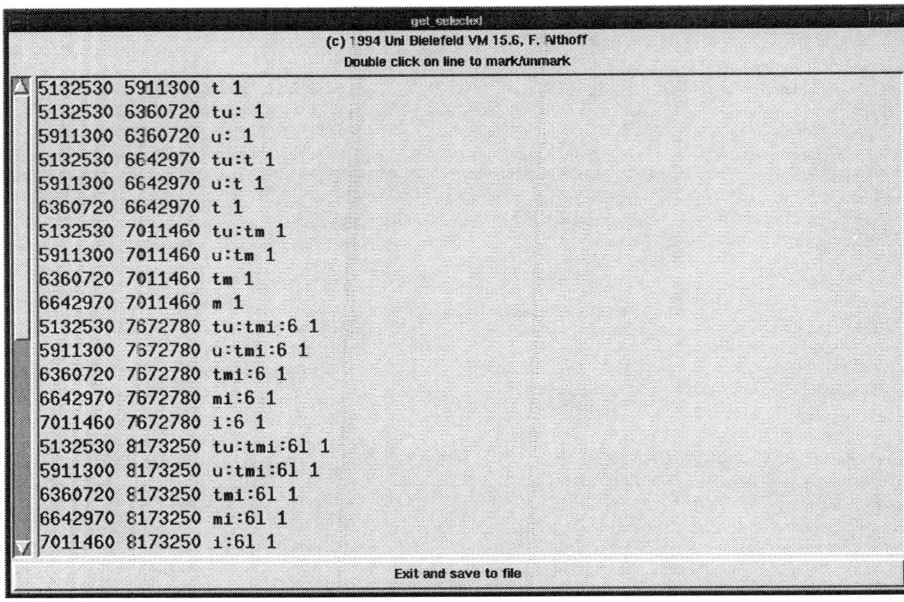

Figure 9.9: Marking Syllable Units for Reference Files

In addition to the generation of reference files, an ESPS label file can be generated for all marked substructures of the utterance on the basis of the format defined above. A direct alignment of units to the signal is therefore possible.

9.5.2 Lexicon Generation

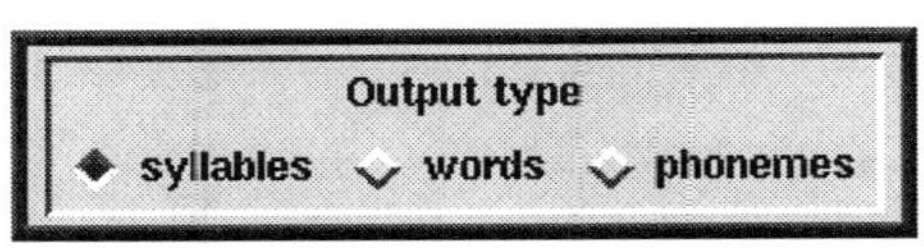

Figure 9.10: Lexicon Generation

GOAL: generate component-specific lexica for the unit

INPUT:

- the unit
- a script listing label files from which a lexicon is to be generated.

OUTPUT: a lexicon the entries of which are in feature (event) notation and include information on frequency of occurrence, average temporal duration of the units and the phonemic transcription.

METHOD: for each unit in the label file generate a lexicon entry by:

- mapping from phonemes to features
- calculate the average temporal duration of the units
- calculate the frequency of the units

The output of lexicon generation is currently component-specific. That is to say, the syllable recognition component has its own internal lexicon format. It is this format which is generated here. However, the information contained in this lexicon could be output in another more general format if required.

9.5.3 Lexicon Consistency Test

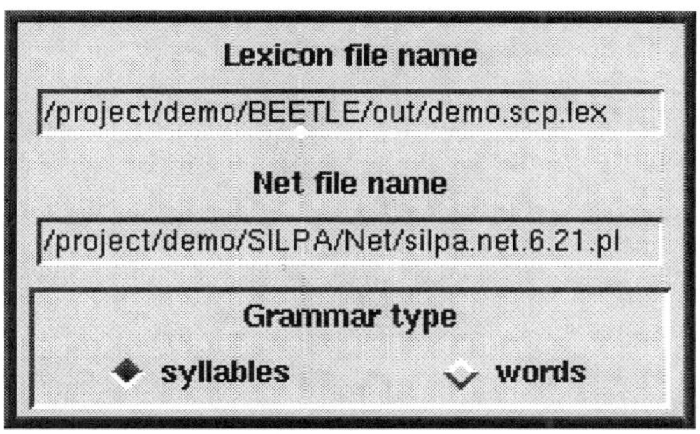

Figure 9.11: Lexicon Consistency Test

GOAL: test the knowledge components for soundness and completeness providing diagnostics.

INPUT: component-specific lexicon and grammar

OUTPUT: the set of lexicon entries which may not be analysed by the grammar.

METHOD: test the generated lexicon (in phonemic format) with the grammar of the component verifying that all lexicon entries may be analysed by the grammar. If the output set is empty, then the lexicon consistency test is successful. If if the grammar does not permit analysis of all lexicon entries, than either a correction of the lexicon or a revision of the grammar is necessary

9.5.4 Top-down Event Generation

The tool used for top-down event generation has been described in detail in Carson-Berndsen (1992c) and Carson-Berndsen (1992d). Here event structures are derived top-down from phonemically labelled data. These structures correspond to optimal labelled input data and form the basis for the data model.

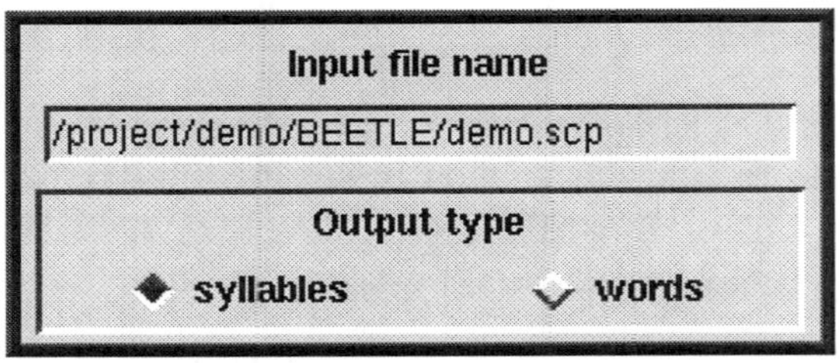

Figure 9.12: Top-down Event Generation

GOAL: generate event structures top-down from phonemically labelled data. These structures correspond to optimal input data and form the basis for the data model

INPUT: a script of phonemic label files for which event representations are to be generated

OUTPUT: event representation files which serve as optimal input files for the component which have a direct alignment to the signal

METHOD: for each label in the input files, lookup is performed which maps the phoneme labels to the relevant events or features and the obligatory contour principle (smoothing) is performed.

9.5.5 Linguistic Word Recognition

GOAL: produce the component-specific output lattices for the evaluation

INPUT: parsing type (i.e. syllables or words)

OUTPUT:

- The components produce lattices in the following format:

  Node-1 Node-2 Label Confidence Start-Time End-Time

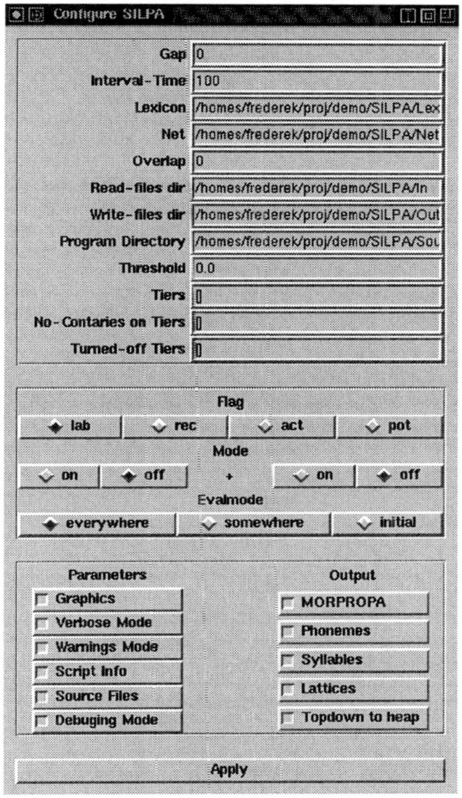

Figure 9.13: SILPA2 Configuration

- The standard output produced by the components is in the following tuple notation:

  ⟨ Syllable—Phoneme, Start-Time, End-Time, Lex-Key ⟩

METHOD: start the parsers by initialising the parameter files.

Lex-Key refers to the type of syllable recognised with respect to the lexicon. *lab* refers to those syllables which are labelled with respect to the corpus, *act* refers to those syllables which are currently in the German language and *pot* refers to those syllables which are well-formed with respect to the phonotactic constraints of German (i.e. new syllables).

9.5.6 String Alignment Evaluation

Within BEETLE, a standard string alignment evaluation procedure using dynamic programming and Levenshtein weights (Lehning (1994)) is called to perform evaluation on the output lattices of SILPA2 and MORPROPA.

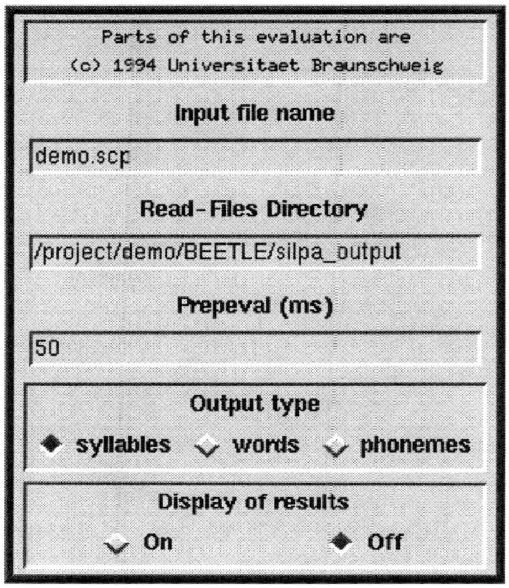

Figure 9.14: String Alignment Evaluation

GOAL: To evaluate nonconnected lattices by using a string alignment evaluation algorithm

INPUT:

- output lattices of the individual components
- the unit
- the unit-specific lexicon

OUTPUT:

- an evaluation document
- a visualisation of the hypothesis graph showing the mapping to the reference string

METHOD:

- map the lattice to a chart analogously to Chien et al. (1990) allowing for a certain amount of overlap (as defined by the parameter *prepeval*.

- perform standard string alignment evaluation procedure using dynamic programming and Levenshtein weights (e.g. Lehning (1994)).
- provide a visualisation of both the input lattice and the output chart using GraphHypo [3], a general tool for displaying hypotheses with respect to either the signal times or the logical nodes of the chart and showing their alignment to the signal.

Table 9.1 presents the translation of a typical evaluation output document generated by the string alignment evaluation developed by Lehning (1994). The original document was in German.

Table 9.1: Example Evaluation Document for String Alignment Evaluation

```
Sentence Identification:  502
Sentence Number: 2

     REF: v a n g e d e6 n E: C S t @ TS u k n a: x
     HYP: v a n g e d e6 n E: C * t @ S  u k n a: x

Error Statistics
   No. of Errors:  2
   Insertions:     0
   Substitutions:  1
   Deletions       1
No. of Hypotheses:    147
No. of Words:         25
Hypotheses per Word:    5.8800

      ========================================
                  Total Results
      ========================================

SENTENCE RECOGNITION RATES
No. of Sentences:  105
   correct:                    0.0% ( 0)
   with Errors:              100.0% ( 105)
      with Substitutions:     81.0% (  85)
      with Deletions:         85.7% (  90)
      with Insertions:        34.3% (  36)

WORD RECOGNITION RATES
   correct              =   44.2%  ( 1824)     39.3/Satz
   Substitutions        =   20.1%  (  830)      7.9/Satz
   Deletions            =   35.8% ( 1477    )  14.1/Satz
   Insertions           =   11.4%  (  472)      4.5/Satz
   Errors (total)       =   67.3%  ( 2779)     26.5/Satz

   Reference Words:                    4131      39.3/Satz
   Hypothesis Words:                   3126      29.8/Satz
   Aligned Words:                      4603      43.8/Satz

INFORMATION ON WORD GRAPH
   No. of Hypotheses:          22157
   Hypotheses per Word:        5.364
```

[3]implemented by Frederek Althoff, 1994

9.5.7 BELLE Evaluation

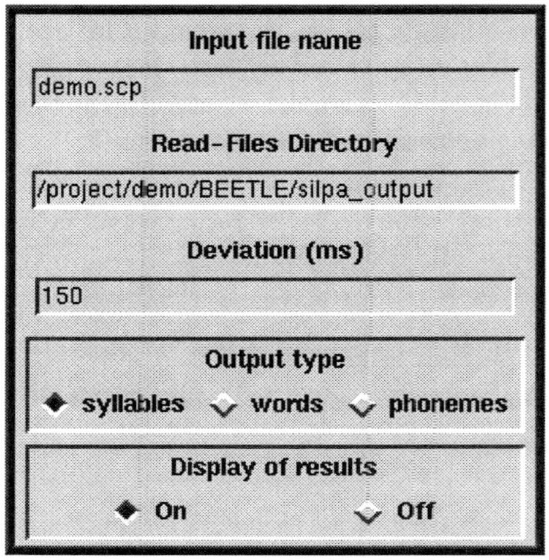

Figure 9.15: BELLE Evaluation

GOAL: To evaluate nonconnected lattices using the temporal distance measure

INPUT:

- nonconnected lattice files
- temporally annotated reference files
- the unit
- a deviation factor

OUTPUT:

- an evaluation document
- a visualisation of the hypothesis graph showing the mapping to the reference string

METHOD: perform the deviation factor evaluation.

The BELLE evaluation procedure has the following characteristics:

- it caters for the notion of overlaps and gaps in the hypothesis lattice.
- it does not assume that a connected path through the lattice exists.

- the reference files used for the evaluation procedure take the temporal annotations of the reference units into account.
- a unit is regarded as recognised if, in the output lattice, there is a corresponding unit with temporal annotations which deviate from the reference unit by not more that the specified deviation parameter
- it only calculates deletions which are defined as a stretch of signal for which no unit matching the reference unit is found.
- it does not currently calculate substitutions and insertions since, in order to be able to apply this information in a linguistically meaningful way, the notion of phonetic similarity should be taken into consideration. Standard evaluation selects the n-best paths through the output lattice with respect to the confidence values provided for each unit. Since SILPA2 utilises underspecified phonological event structures, it implicitly defines the notion of phonetic or phonological similarity between units. It would seem more suitable, instead of selecting the *best* hypothesis (w.r.t some confidence value) at a particular point in time, to select the hypothesis which is the nearest, phonologically speaking, to the reference unit. Since the syllable and phoneme hypotheses are generated by *multiplying out* the underspecified phonological event structures, the corresponding fully specified event structures are more closely related phonologically than other fully specified event structures which are not subsumed by this.
- the recognition results are defined solely on the basis of the percentage of the correctly matched units with respect to the total number of reference units.
- the hypothesis density (the total number of hypotheses with respect to the number of reference units) provides some indication of the accuracy of the system.

Table 9.2 shows an example evaluation document produced by the BELLE evaluation.

Figure 9.16 shows a visualisation of the output of the BELLE evaluation procedure for a syllable lattice. The shaded area represents hypotheses found in the output lattice which correspond to the reference file for this utterance. The nonshaded hypotheses represent the reference path as generated by Reference File Generation. Hypotheses which do not correspond to the reference path are not shown, for reasons of clarity. As can be seen from the figure, there is an overlap of nasality in the combination /fɔnmʏn/ in *von München.* These syllables would not be regarded as recognised in the string alignment procedure without an arbitrary splitting into two nasal segments. However, since the

Table 9.2: Example Evaluation Document for BELLE Evaluation

```
          Linguistic Word Recognition Evaluation for syllables

Date: Tue Jun  4 11:07:40 1996

Institution:  LILI/UBI

Deviation Factor: 150

          ------------------------------------------

Reference File:  demo.syl.ref
Hypothesis File: demo


REF: tu:t  mi:6  laIt
HYP: tu:t  mi:6  laIt


No. of Correct    3
No. of Gaps       0
No. of Hypotheses 105
No. of Units      3


Recognition Rate  100.00%
Hypothesis Density 35.00

         ====================================

               TOTAL RECOGNITION RATE

         ====================================

RECOGNITION RATE   100.00%
No. of Utterances  1
Hypothesis Density 35.00

         (c) 1994 University of Bielefeld, LiLi
```

place of articulation is underspecified in the input to the phonological parser, an arbitrary splitting of such a nasal segment into two further segments is not justified.

The GRAPHHYPO visualisation tool has been parametrised to allow either all hypotheses to be displayed or only the hypotheses which match the reference units. The distinction is shown in Figures 9.17 and 9.18. The reference path is displayed next to the signal in each case. In Figure 9.17 the darker shading represents the hypotheses which do not match the reference units, the lighter shading represent those that do. In Figure 9.18 only the hypotheses which match the reference units are shown. Note that no smoothing over temporally similar hypotheses has been undertaken and therefore there may be a number of matching hypotheses for each reference unit.

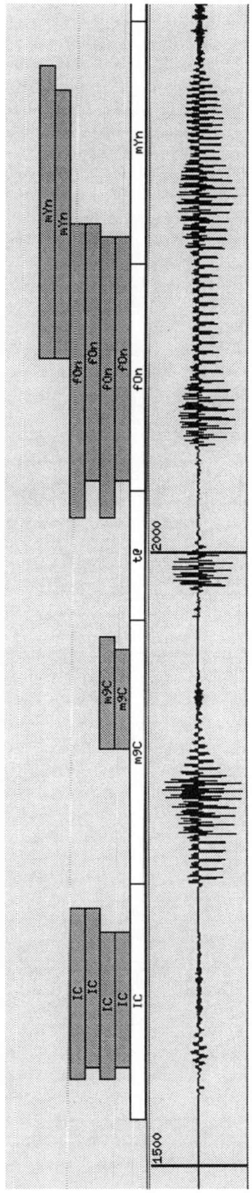

Figure 9.16: Visualisation

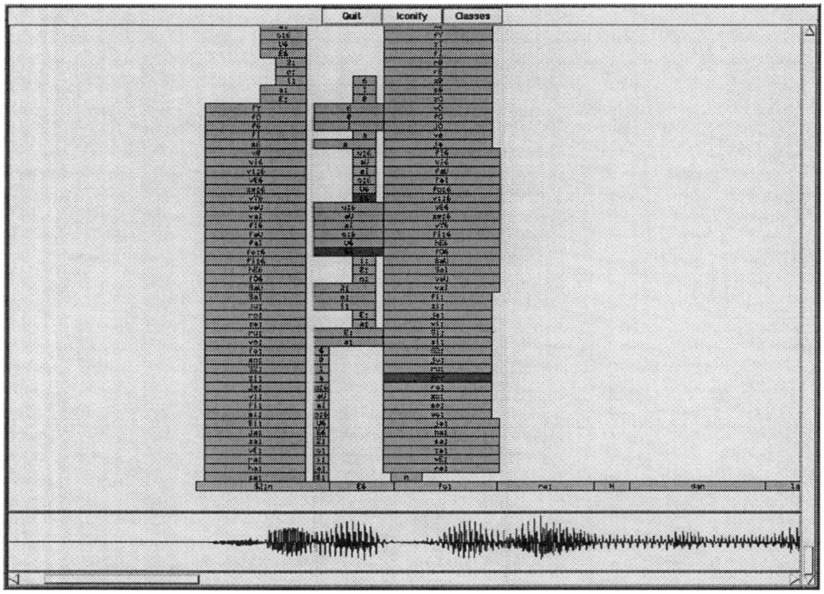

Figure 9.17: GRAPHHYPO visualisation of all hypotheses

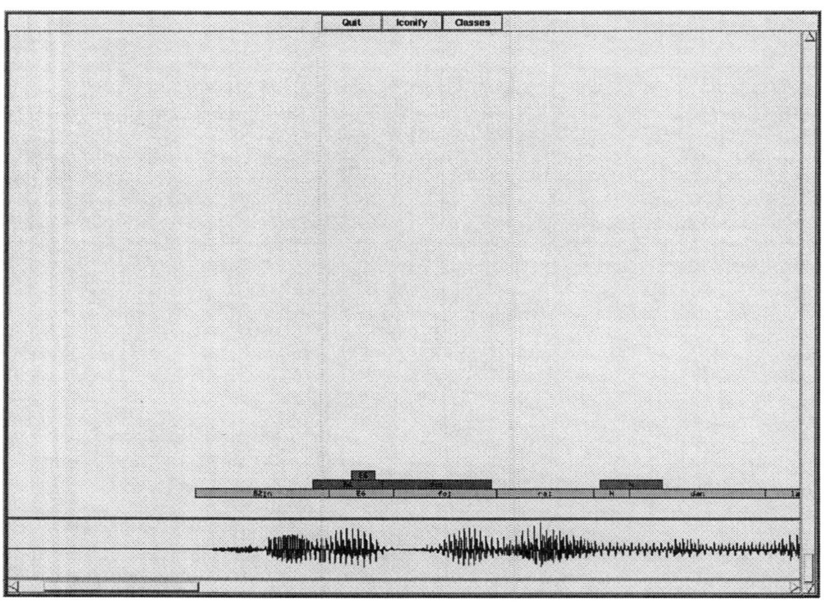

Figure 9.18: GRAPHHYPO visualisation of only matching hypotheses

This conclues the discussion of the BEETLE toolkit. The architecture is shown in figure 9.19.

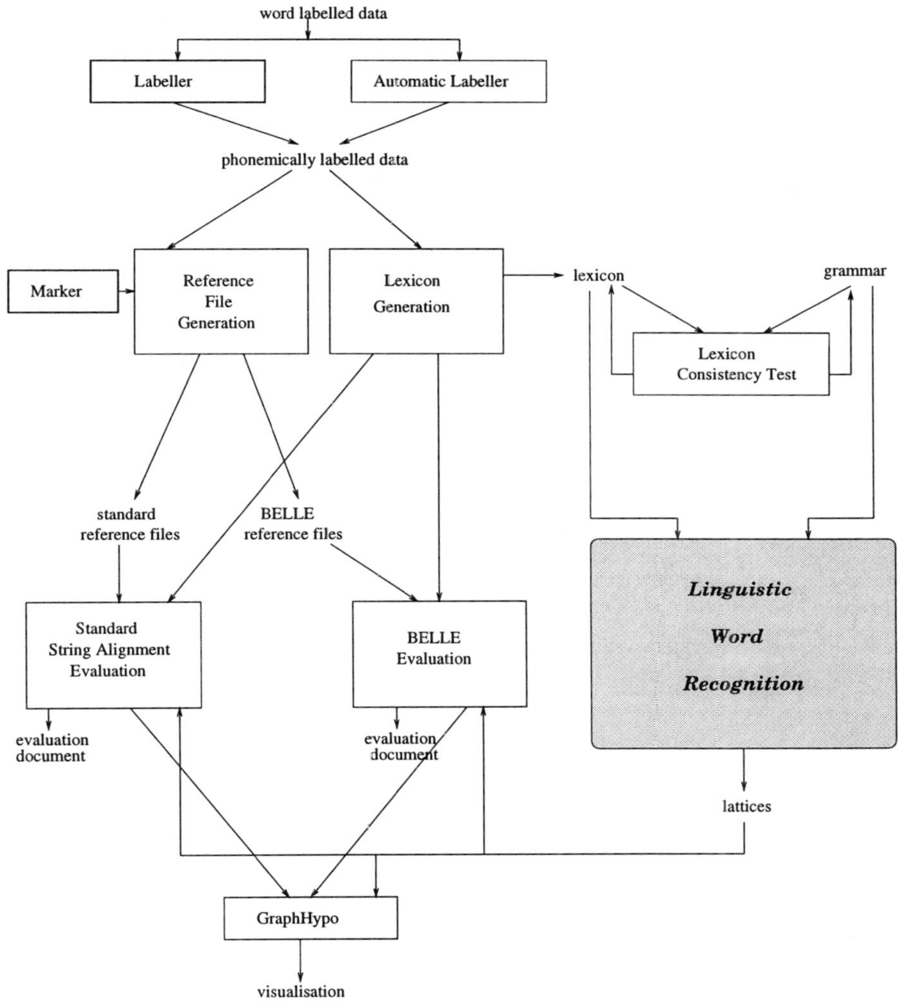

Figure 9.19: The BEETLE Architecture

9.6 EVALUATION RESULTS

The event-based phonological parser, SILPA2, was evaluated according to the stages of diagnostic evaluation defined in this chapter. Logical evaluation was carried out with respect to two data models corresponding to the scenario data specified in Table 9.3. Logical recognition rates of 98.9% and 99.2% respectively were attained. The error rate in each case was due to the fact that no phonologically relevant temporal statistics were calculated. Empirical evaluation was carried out using the output of the HEAP acoustic event recogniser as input to SILPA2. Here again, two data sets were evaluated.

Table 9.3: Data Set Specifications

Data Set	Speaker Type	Scenario Type	No. of Utterances Evaluated
A	single speaker	railway information	200
B	many speaker	scheduling task dialogues	82

For empirical evaluation of SILPA2 a number of parametrisations were selected. They are defined in Table 9.4 and refer to the definition of constraint types discussed in Section 8.4. OPCs refer to *obligatory performance constraints*, NCCs refer to *No Contrary Constraints* and UPCs refer to *Unreliability Performance Constraints*. Each of these constraints specify phonological tiers in the description. Overlap and Gap refer to the temporal parameter settings and Top-Down Constraints refer to whether top-down constraints were sent to the acoustic event recogniser HEAP.

Table 9.4: Parameter Set Specifications for SILPA2

Parameter Set	OPCs	NCCs	UPCs	Overlap	Gap	Top-Down Constraints
P1	manner	place	{}	50	50	off
P2	{}	{}	{}	50	50	off
P3	{}	{}	{}	50	50	on

The initial evaluation of SILPA2 was done on the railway information scenario using 200 utterances of a single speaker. Two parametrisations of the system were tested giving the results displayed in Table 9.5. As can be seen from the table, significant improvements can be attained by evoking strong constraint relaxation with the parameter specification P2.

Table 9.5: Evaluation Results: Parameter Set A

Data Set	Parametrisation	Phoneme Recognition Rate	Syllable Recognition Rate
A	P1	49.6%	14.5%
A	P2	72.5%	37%

The second empirical evaluation of SILPA2, again using the output of the

HEAP acoustic event recogniser as input to SILPA2, was carried out on 82 utterances of the scheduling tasks scenario. The results are displayed in Table 9.6. The hypothesis density is given in brackets after the phoneme and syllable recognition rates in each case. The relatively high hypothesis density is a result of the fact that the standard output from SILPA2 is in the form of underspecified complex phonological event structures; these structures are then 'multiplied out' into their full phonemic forms for evaluation purposes, resulting in several forms for a single underspecified structure. The hypothesis density does not therefore mirror the hypothesis density which is relevant for the interface between SILPA2 and MORPROPA.

In this case again, improvements were made by introducing strong constraint relaxation (P2). The parametrisation specification P3 refers to the very first evaluation of top-down interaction between SILPA2 and HEAP. The strategy taken by HEAP in evaluating top-down predictions was to "turn off" all detectors for events which had not been predicted. This did not lead to any significant changes in the evaluation results. Other strategies are currently work in progress and therefore no evaluation results are available at present.

The results are remarkable for a purely knowledge-based system. It is anticipated that enhancement by stochastic information will lead to further improvement of the recognition results.

Table 9.6: Evaluation Results: Parameter Set B

Data Set	Parametrisation	Phoneme Recognition Rate	Syllable Recognition Rate
B	P1	54% (18.25)	21% (31.75)
B	P2	66.97% (19.22)	35.19% (58.77)
B	P3	66.65% (19.15)	34.71% (58.74)

10

CONCLUSION

The goal of this book was to present an empirically and representationally adequate computational model of phonology for processing in a spoken language recognition system. A new constraint-based temporal interpretation approach to phonological parsing was presented which, starting from recent well-motivated developments in multi-linear phonology, in particular autosegmental phonology and phonological events, offers a principled and elegant solution to the projection problem in spoken language recognition.

Time Map Phonology, a temporal interpretation approach to phonological description takes autosegmental phonology as the descriptive model and utilises a flexible notion of compositionality based on underspecified structures with 'autosegmental' tiers of parallel phonetic and phonological events. A finite state automaton was used as the implementation model for the event logic (cf. Kasper & Rounds, 1986) and empirical evidence was drawn from spoken language recognition and experimental phonetics. This approach avoids the necessity in spoken language recognition for an early rigid mapping from phonetic parameters to simple sequences of segments. It was shown that, in this way, all information in the acoustic signal can be made available to phonological processing and, if required, to higher levels of processing. By allowing a representation in terms of overlapping events at the phonetics/phonology interface, a representation which is nearer to phonetics than the phoneme sequences of more traditional phonologies and concatenative speech recognition models could be obtained. This approach goes demonstrably further than previous formal interpretations of autosegmental phonology in that phonological relations are explicitly related to intervals in actual speech signals as required by a spoken language recognition system. It was illustrated that by distinguishing between specified and unspecified temporal constraints on phonetic and phonological representations and by the incorporation of constraint relaxation and constraint enhancement mechanisms, the parsing process can be made more robust.

There are, however, a number of issues which require further consideration. These concern the empirical adequacy of complex phonetic event descriptions as mentioned in Chapter 5, the realisation of appropriate goal-directed constraint relaxation and constraint enhancement mechanisms in the parsing process, the treatment of confidence values during processing and the introduction of explicit temporal statistics. Each of these issues is discussed in the next section.

Although possible solutions to these open questions have been considered as part of the accompanying research for this book, these issues are still work in progress. Support tools for designing heuristics for use in solving a number of these open questions have been developed. They are discussed briefly in Section 10.2.

10.1 OPEN ISSUES

There are a number of issues which are currently being considered for the further development of the event-based approach to phonetic and phonological parsing. The first issue which is considered further here concerns the simplex and complex phonetic events described in Chapters 4 and 5. Complex phonetic events are described by networks of temporal constraints on acoustic events (simplex phonetic events).The complex phonetic events have an articulatory character in the individual network definitions. This is not a necessity. They are defined in terms of articulatory features in this book because these can easily be defined top-down on the basis of phonological descriptions. This allows for a clear distinction between types of events in the acoustic, phonetic and phonological knowledge domains. The phonetic knowledge domain can refer to both acoustic and articulatory events and the phonological knowledge domain can refer to articulatory and other structural events which are not classified as articulatory (e.g. onset events, syllable events). It will be seen in connection with the discussion in the next section that if tools for the automatic construction of such networks are applied in this area, then there is no reason why other event-feature classifications should not be employed.

More research is required before the complex phonetic event networks can be deemed to be empirically adequate. The networks as currently described in Chapter 5 have been constructed on the basis of relatively limited data on acoustic events. Note that this does not apply to the phonotactic network which has been constructed and fully tested according to the phonotactic rules of the language. The finite-state automaton model of the network representation formalism allows for an elegant mechanism for empirical testing of the complex phonetic networks.

Another issue which has not been approached in connection with the event-based approach to phonological parsing presented in Chapter 7 concerns the treatment of hypothesis priorities (confidence values) in the individual knowledge domains. Such hypothesis weighting will lead to a reduction of the local search space at each level of analysis. However, much empirical testing has to be done in this connection and in particular with respect to the treatment of top-down predictions. Research in connection with BELLEx3.2 has shown that a strategy which allows its search space to be totally confined by top-down predictions is too restrictive. A more appropriate strategy is to introduce a threshold setting for local hypotheses allowing local hypotheses with a high confidence value to have an equal standing with top-down hypotheses. The network representation of the knowledge component in the phonetic and phonological knowledge domains allows for a possible arc weighting (or ordering) which could be an initial step towards defining local confidence values. An example of this would be the weighted representation of syllable onset maximisation using the notion of early-closure (Kimball, 1973) by weighting termination at a given final node higher than continuation.

Another issue which has not been treated in the discussion above is the introduction of temporal statistics (e.g. duration, percentage overlap) into the knowledge components. This would reduce the number of possible hypotheses. Explicit temporal statistics have not been incorporated into the system so far. As was seen above, constraints are on temporal relations and do not contain any further temporal information. This is also work in progress. Tools which have been developed for this purpose are discussed in Section 10.2. However, temporal statistics as heuristics for processing are regarded as an additional aid and not as a stochastic model. As was seen in Chapter 7, the model works without such temporal statistics, but the search space can be reduced considerably by incorporating them as stochastic heuristics.

10.2 TOOLS FOR DESIGNING HEURISTICS

In order to design heuristics and gather information for solving each of the open issues discussed above, tools have been developed which are sketched briefly here. The first tool or rather toolkit concerns the definition of the complex event networks in all knowledge domains, and uses an approach similar to the automata induction of Berwick & Pilato (1987) for constructing networks according to a primary tier. The initial work on this corresponded to the different types of network representations for phonotactics described in Chapter 5 (in particular Section 5.3). This work involved the automata learning of a phoneme-based syllable network based on positive examples and then the use of additional tools for the construction of natural classes of features, of CV-tiers and of redundancy (feature cooccurrence) rules. The details of these tools are not discussed further here: it would be possible (and perhaps more appropriate) to devote a further book to the possibilities of learning phonological descriptions.

Knowledge acquisition tools for collecting temporal statistics from phonemic label files, for generating phonological event structures on the basis of phonemic transcriptions and for generating phonological event label files have been also developed for event phonologies. These tools are described in detail in Carson-Berndsen (1992c, 1992d). Using these tools, and starting with a phonemic label file, it is possible to generate a phonological event label file which has a direct correspondence to the speech signal. The phonological event label files can then be reanalysed interactively using speech signal labelling software (e.g. SAMLAB, Braun, 1991a) and where possible the interval endpoints altered or corrected by trained labelling personnel. Furthermore, the individual phonological event label files may together be employed in multi-tier prosodic labelling systems to provide top-down knowledge for the speech signal labelling software. These tools have also been applied to simulate input to the phonological parser for testing purposes during development.

Using the temporal annotations for phonological events constructed by the event label file generation tool and the temporally annotated acoustic events from other sources, another tool was developed which, for a token utterance, compares and contrasts the acoustic and phonological events and calculates the percentage overlap between events at each level and also constructs for each event at each level, the concurrent events at the other level and the temporal relations between them. A graphic output of the phonological and acoustic events is also provided for manual comparison and verification. This tool is currently being applied in the design of a heuristic which decides whether two events participate in an overlap or a precedence relationship. This problem was discussed in Chapter 7 in connection with the axioms governing absolute-time events. The current solution is to assume both relationships, except where overlap was also temporal inclusion. The temporal statistics on percentage overlap will therefore play a considerable role in the reduction of the search space for event relations.

All the tools mentioned in this section contribute important information which will aid in the design of efficient constraint relaxation and constraint enhancement mechanisms. Furthermore tools for testing these mechanisms could be specified which take speaker performance criteria in different phonostylistic contexts into account.

An interesting further perspective for the present declarative approach would be the definition of an "inverse" procedural semantics for speech synthesis; this is an open task which has, however, been foreshadowed to some extent by the event label file generation tool described above.

10.3 FINAL REMARKS

In the above sections, it was sketched how the declarative approach taken in this book can be developed further. However, all of these further developments concern robustness and the reduction of the search space during processing. As it stands, the approach taken here offers a novel alternative to the concatenative models of current spoken language recognition systems with broader coverage and more flexibility than these, in principle, in terms of the projection problem of new words and speech variation. Although the input to the system was limited to elementary acoustic events, there is no empirical restriction made on what "events" can serve as input to the system. Demisyllables, for example, or tone, intensity, and rhythmic ("metrical") information could provide additional sources of information. Indeed the relationship of *Time Map Phonology* to Metrical Phonology has yet to be explored further. Although *Time Map Phonology* does distinguish between syllable types and syllable-part events which play a role in metrical structure, the connection to such a level has still to be made explicit.

An issue which has not yet been considered in connection with the work described here is computational complexity. This is an issue with which finite-state phonologies have had to address (cf. the criticisms of two-level morphology in Barton et al. (1987) and the discussions in Koskenniemi & Church (1988) and Ritchie et al. (1992)). Due to the fact that the event-based model can be interpreted by a deterministic finite state automaton (as opposed to the finite state transducers with ϵ-transitions of two-level morphology) which uses underspecification rather than disjunction, it is conjectured here that it is not NP-hard. The questions of algebraic complexity and system performance was not made an issue here. This is question for future consideration, whenever, the extensions described in the previous sections have been incorporated into the system.

APPENDIX A

SAMPA PHONETIC ALPHABET FOR GERMAN

Consonants

p	[paIn]	Pein
b	[baIn]	Bein
t	[tot]	Tod
d	[daIn]	dein
k	[kaIn]	kein
g	[gut]	gut
m	[maIn]	mein
n	[naIn]	nein
N	[laN]	lang
f	[faIn]	fein
v	[vaIn]	Wein
s	[las]	laß
z	[zaIn]	sein
S	[SaIn]	Schein
Z	[Ze$ni]	Genie
C	[IC]	ich
x	[ax]	ach
R	[RaIn]	rein
h	[haIn]	Hain
j	[ju$b@l]	Jubel
l	[laUt]	Laut

Vowels

i	[bi$t@n]	bieten
I	[bIt@n]	bitten
y	[my$d]	müde
Y	[bY$t@n]	Bütten
e	[be$t@n]	beten
2	[b2$t@n]	böten
9	[b9$k@]	Böcke
E:	[bE:$t@n]	bäten
E	[bE$t@n]	Betten
u	[bu$d@n]	Buden
U	[bU$t6]	Butter
o	[bo$t@n]	Boten
O	[mO$t@n]	Motten
a:	[ba:$t@n]	baten
a	[ma$t@n]	Matten
@	[bIt@]	bitte
6	[bI$t6]	bitter

Affricates

pf	[pfa:l]	Pfahl
ts	[tsa:l]	Zahl
tS	[matS]	Matsch

Diphthongs

aI	[laI$z@]	leise
aU	[laUs]	Laus
OY	[lOY$z@]	Läuse

Boundaries

$	Syllable Boundary

Diacritics

H	Aspiration
=	Syllabic

APPENDIX B

THE CHOMSKY HIERARCHY

A grammar can be classified according to its production rules into regular grammars, context free grammars, context sensitive grammars and unrestricted rewriting systems. The notion of *hierarchy* is expressed in the fact that: every Type 3 grammar is also a Type 2 grammar; Type 3 and Type 2 grammars are also Type 1 grammars, and Type 3, Type 2 and Type 1 grammars are also Type 0 grammars. Similarly a language which can be characterised by a Type 1 grammar can also be characterised by a Type 3 grammar but not vice versa. These facts are sketched briefly in the table below. The equivalence between the grammar types and formal devices is also given. A detailed account of these issues can be found in Partee et.al (1990: 451f) and Hopcroft & Ullman (1979: 217ff). The following rule types exist:

Type 0: have rules of the form $\alpha \rightarrow \beta$ where α and β are arbitrary strings where $\alpha \neq \epsilon$. Such grammars are termed *unrestricted rewriting systems*. They are equivalent to Turing Machines

Type 1: have rules of the form $\alpha \mathrm{A} \beta \rightarrow \alpha \mathrm{B} \beta$ where $\mathrm{B} \neq \epsilon$. Such grammars are termed *context sensitive* grammars. A language is context sensitive if and only if it can be characterised by a context sensitive grammar which is not also context free. Context sensitive grammars are equivalent to linear bounded automata.

Type 2: have rules of the form $\mathrm{A} \rightarrow \mathrm{BC}$ or $\mathrm{A} \rightarrow x$. Such grammars are termed *context free* grammars. A language is context free if and only if it can be characterised by a context free non-regular grammar. Context free grammars are equivalent to pushdown automata.

Type 3: have rules of the form $\mathrm{A} \rightarrow x\mathrm{B}$ or $\mathrm{A} \rightarrow x$. Such grammars are termed *regular* or *linear* grammars. A language is regular if and only if it can be characterised by a regular grammar. Regular grammars are equivalent to finite state automata.

APPENDIX C

EVENT-BASED PHONOTACTIC NETWORK FOR GERMAN

Note that r is currently restricted to the consonantal variant /R/, a voiced uvular fricative.

$$\begin{array}{l}\langle 0,\\ \quad \{\langle I_{ph}^{0}, \langle \text{PHONATION:voiceless}\rangle\rangle, \langle I_{ma}^{0}, \langle \text{MANNER:fricative}\rangle\rangle, \langle I_{pl}^{0}, \langle \text{PLACE: palato-alveolar}\rangle\rangle\},\\ \qquad \{\langle I_{ph}^{0}, {}^{\circ}, I_{ma}^{0}\rangle, \langle I_{ma}^{0}, {}^{\circ}, I_{pl}^{0}\rangle, \langle I_{ph}^{0}, {}^{\circ}, I_{pl}^{0}\rangle\},\\ \qquad\quad \{\langle I_{ph}^{0}, \prec^{\circ}, I_{ph}^{1}\rangle, \langle I_{ma}^{0}, \prec^{\circ}, I_{ma}^{1}\rangle, \langle I_{pl}^{0}, \prec^{\circ}, I_{pl}^{1}\rangle\},\\ \qquad\quad\; 1\rangle\end{array}$$

$$\begin{array}{l}\langle 0,\\ \quad \{\langle I_{ph}^{0}, \langle \text{PHONATION:voiceless}\rangle\rangle, \langle I_{ma}^{0}, \langle \text{MANNER:fricative}\rangle\rangle, \langle I_{pl}^{0}, \langle \text{PLACE: apical}\rangle\rangle\},\\ \qquad \{\langle I_{ph}^{0}, {}^{\circ}, I_{ma}^{0}\rangle, \langle I_{ma}^{0}, {}^{\circ}, I_{pl}^{0}\rangle, \langle I_{ph}^{0}, {}^{\circ}, I_{pl}^{0}\rangle\},\\ \qquad\quad \{\langle I_{ph}^{0}, \prec^{\circ}, I_{ph}^{2}\rangle, \langle I_{ma}^{0}, \prec^{\circ}, I_{ma}^{2}\rangle, \langle I_{pl}^{0}, \prec^{\circ}, I_{pl}^{2}\rangle\},\\ \qquad\quad\; 2\rangle\end{array}$$

$$\begin{array}{l}\langle 0,\\ \quad \{\langle I_{ph}^{0}, \langle \text{PHONATION:voiceless}\rangle\rangle, \langle I_{ma}^{0}, \langle \text{MANNER:affricate}\rangle\rangle, \langle I_{pl}^{0}, \langle \text{PLACE: apical}\rangle\rangle\},\\ \qquad \{\langle I_{ph}^{0}, {}^{\circ}, I_{ma}^{0}\rangle, \langle I_{ma}^{0}, {}^{\circ}, I_{pl}^{0}\rangle, \langle I_{ph}^{0}, {}^{\circ}, I_{pl}^{0}\rangle\},\\ \qquad\quad \{\langle I_{ph}^{0}, \prec^{\circ}, I_{ma}^{3}\rangle, \langle I_{ma}^{0}, \prec^{\circ}, I_{ma}^{3}\rangle, \langle I_{pl}^{0}, \prec^{\circ}, I_{pl}^{3}\rangle\},\\ \qquad\quad\; 3\rangle\end{array}$$

$$\begin{array}{l}\langle 0,\\ \quad \{\langle I_{ma}^{0}, \langle \text{MANNER:plosive}\rangle\rangle, \langle I_{pl}^{0}, \langle \text{PLACE: } \alpha\rangle\rangle\} \mid \alpha \in \{\text{labial, velar}\},\\ \qquad \{\langle I_{ma}^{0}, {}^{\circ}, I_{pl}^{0}\rangle\},\\ \qquad\quad \{\langle I_{ma}^{0}, \prec^{\circ}, I_{ma}^{4}\rangle, \langle I_{pl}^{0}, \prec^{\circ}, I_{pl}^{4}\rangle\},\\ \qquad\quad\; 4\rangle\end{array}$$

$$\begin{array}{l}\langle 0,\\ \quad \{\langle I_{ph}^{0}, \langle \text{PHONATION:voiceless}\rangle\rangle, \langle I_{ma}^{0}, \langle \text{MANNER:fricative}\rangle\rangle, \langle I_{pl}^{0}, \langle \text{PLACE: } \alpha\rangle\rangle\} \mid \alpha \in \{\text{palato,labial}\},\\ \qquad \{\langle I_{ph}^{0}, {}^{\circ}, I_{ma}^{0}\rangle, \langle I_{ma}^{0}, {}^{\circ}, I_{pl}^{0}\rangle, \langle I_{ph}^{0}, {}^{\circ}, I_{pl}^{0}\rangle\},\\ \qquad\quad \{\langle I_{ph}^{0}, \prec^{\circ}, I_{ph}^{4}\rangle, \langle I_{ma}^{0}, \prec^{\circ}, I_{ma}^{4}\rangle, \langle I_{pl}^{0}, \prec^{\circ}, I_{pl}^{4}\rangle\},\\ \qquad\quad\; 4\rangle\end{array}$$

$$\begin{array}{l}\langle 0,\\ \quad \{\langle I_{ph}^{0}, \langle \text{PHONATION:voiceless}\rangle\rangle, \langle I_{ma}^{0}, \langle \text{MANNER:affricate}\rangle\rangle, \langle I_{pl}^{0}, \langle \text{PLACE: labial}\rangle\rangle\},\\ \qquad \{\langle I_{ph}^{0}, {}^{\circ}, I_{ma}^{0}\rangle, \langle I_{ma}^{0}, {}^{\circ}, I_{pl}^{0}\rangle, \langle I_{ph}^{0}, {}^{\circ}, I_{pl}^{0}\rangle\},\\ \qquad\quad \{\langle I_{ph}^{0}, \prec^{\circ}, I_{ph}^{4}\rangle, \langle I_{ma}^{0}, \prec^{\circ}, I_{ma}^{4}\rangle, \langle I_{pl}^{0}, \prec^{\circ}, I_{pl}^{4}\rangle\},\\ \qquad\quad\; 4\rangle\end{array}$$

$<0,$
$\{<I_{ma}{}^{0}, <\text{MANNER:plosive}>>\},$
$\{\},$
$\{<I_{ma}{}^{0},\prec^{\circ},I_{ma}{}^{5}>\},$
$5>$

$<0,$
$\{<I_{ph}{}^{0}, <\text{PHONATION:voiceless}>>, <I_{ma}{}^{0}, <\text{MANNER:fricative}>>, <I_{pl}{}^{0}, <\text{PLACE:}\ \alpha>>\} \mid \alpha \in \{\text{labial, palatal}\},$
$\{<I_{ph}{}^{0},^{\circ},I_{ma}{}^{0}>, <I_{ma}{}^{0},^{\circ},I_{pl}{}^{0}>, <I_{ph}{}^{0},^{\circ},I_{pl}{}^{0}>\},$
$\{<I_{ph}{}^{0},\prec^{\circ},I_{ph}{}^{5}>, <I_{ma}{}^{0},\prec^{\circ},I_{ma}{}^{5}>, <I_{pl}{}^{0},\prec^{\circ},I_{pl}{}^{5}>\},$
$5>$

$<0,$
$\{<I_{ph}{}^{0}, <\text{PHONATION:voiceless}, <I_{ma}{}^{0}, <\text{MANNER:affricate}>>, <I_{pl}{}^{0}, <\text{PLACE: labial}>>\},$
$\{<I_{ph}{}^{0},^{\circ},I_{ma}{}^{0}>, <I_{ma}{}^{0},^{\circ},I_{pl}{}^{0}>, <I_{ph}{}^{0},^{\circ},I_{pl}{}^{0}>\},$
$\{<I_{ph}{}^{0},\prec^{\circ},I_{ph}{}^{5}>, <I_{ma}{}^{0},\prec^{\circ},I_{ma}{}^{5}>, <I_{pl}{}^{0},\prec^{\circ},I_{pl}{}^{5}>\},$
$5>$

$<0,$
$\{<I_{ph}{}^{0}, <\text{PHONATION:voiced}>>, <I_{ma}{}^{0}, <\text{MANNER:fricative}>>, <I_{pl}{}^{0}, <\text{PLACE: labial}>>\},$
$\{<I_{ph}{}^{0},^{\circ},I_{ma}{}^{0}>, <I_{ma}{}^{0},^{\circ},I_{pl}{}^{0}>, <I_{ph}{}^{0},^{\circ},I_{pl}{}^{0}>\},$
$\{<I_{ph}{}^{0},\prec^{\circ},I_{ph}{}^{5}>, <I_{ma}{}^{0},\prec^{\circ},I_{ma}{}^{5}>, <I_{pl}{}^{0},\prec^{\circ},I_{pl}{}^{5}>\},$
$5>$

$<0,$
$\{<I_{ma}{}^{0}, <\text{MANNER:plosive}>>, <I_{pl}{}^{0}, <\text{PLACE: velar}>>\},$
$\{<I_{ma}{}^{0},^{\circ},I_{pl}{}^{0}>\},$
$\{<I_{ma}{}^{0},\prec^{\circ},I_{ma}{}^{6}>, <I_{pl}{}^{0},\prec^{\circ},I_{pl}{}^{6}>\},$
$6>$

$<0,$
$\{<I_{ph}{}^{0}, <\text{PHONATION:voiceless}>>, <I_{ma}{}^{0}, <\text{MANNER:plosive}>>, <I_{pl}{}^{0}, <\text{PLACE:}\ \alpha>>\} \mid \alpha \in \{\text{labial, velar}\},$
$\{<I_{ph}{}^{0},^{\circ},I_{ma}{}^{0}>, <I_{ma}{}^{0},^{\circ},I_{pl}{}^{0}>, <I_{ph}{}^{0},^{\circ},I_{pl}{}^{0}>\},$
$\{<I_{ph}{}^{0},\prec^{\circ},I_{ph}{}^{6}>, <I_{ma}{}^{0},\prec^{\circ},I_{ma}{}^{6}>, <I_{pl}{}^{0},\prec^{\circ},I_{pl}{}^{6}>\},$
$6>$

$<0,$
$\{<I_{ma}{}^{0}, <\text{MANNER:plosive}>>\},$
$\{\},$
$\{<I_{ma}{}^{0},\prec^{\circ},I_{ma}{}^{peak}>\},$
$\text{peak}>$

$<0,$
$\{<I_{ma}{}^{0}, <\text{MANNER:affricate}>>\text{manner}\},$
$\{\},$
$\{<I_{ma}{}^{0},\prec^{\circ},I_{ma}{}^{peak}>\},$
$\text{peak}>$

$<0,$
$\{<I_{ph}{}^{0}, <\text{PHONATION:voiceless}>>, <I_{ma}{}^{0}, <\text{MANNER:fricative}>>, <I_{pl}{}^{0}, <\text{PLACE:}\ \alpha>>\} \mid \alpha \in \{\text{labial, palatal, glottal, palato}\},$
$\{<I_{ph}{}^{0},^{\circ},I_{ma}{}^{0}>, <I_{ma}{}^{0},^{\circ},I_{pl}{}^{0}>, <I_{ph}{}^{0},^{\circ},I_{pl}{}^{0}>\},$
$\{<I_{ma}{}^{0},\prec^{\circ},I_{ma}{}^{peak}>, <I_{pl}{}^{0},\prec^{\circ},I_{pl}{}^{peak}>\},$
$\text{peak}>$

$<0,$
$\{<I_{ph}^{0}, <\text{PHONATION:voiced}>>, <I_{ma}^{0}, <\text{MANNER:fricative}>>\},$
$\{<I_{ph}^{0}, \circ, I_{ma}^{0}>\},$
$\{<I_{ma}^{0}, \prec^{\circ}, I_{ma}^{peak}>\},$
$\text{peak}>$

$<0,$
$\{<I_{ph}^{0}, <\text{PHONATION:voiced}>>, <I_{ma}^{0}, <\text{MANNER:lateral}>>\},$
$\{<I_{ph}^{0}, \circ, I_{ma}^{0}>\},$
$\{<I_{ma}^{0}, \prec^{\circ}, I_{ma}^{peak}>\},$
$\text{peak}>$

$<0,$
$\{<I_{ph}^{0}, <\text{PHONATION:voiced}>>, <I_{ma}^{0}, <\text{MANNER:fricative}>>, <I_{pl}^{0}, <\text{PLACE: uvular}>>\},$
$\{<I_{ph}^{0}, \circ, I_{ma}^{0}>, <I_{ma}^{0}, \circ, I_{pl}^{0}>, <I_{ph}^{0}, \circ, I_{pl}^{0}>\},$
$\{<I_{ma}^{0}, \prec^{\circ}, I_{ma}^{peak}>, <I_{pl}^{0}, \prec^{\circ}, I_{pl}^{peak}>\},$
$\text{peak}>$

$<1,$
$\{<I_{ph}^{1}, <\text{PHONATION:voiceless}>>, <I_{ma}^{1}, <\text{MANNER:plosive}>>, <I_{pl}^{1}, <\text{PLACE: } \alpha>>\} \mid \alpha \in \{\text{labial, velar}\},$
$\{<I_{ph}^{1}, \circ, I_{ma}^{1}>, <I_{ma}^{1}, \circ, I_{pl}^{1}>, <I_{ph}^{1}, \circ, I_{pl}^{1}>\},$
$\{<I_{ph}^{1}, \prec^{\circ}, I_{ph}^{8}>, <I_{ma}^{1}, \prec^{\circ}, I_{ma}^{8}>, <I_{pl}^{1}, \prec^{\circ}, I_{pl}^{8}>\},$
$8>$

$<1,$
$\{<I_{ph}^{1}, <\text{PHONATION:voiceless}>>, <I_{ma}^{1}, <\text{MANNER:plosive}>>, <I_{pl}^{1}, <\text{PLACE: apical}>>\},$
$\{<I_{ph}^{1}, \circ, I_{ma}^{1}>, <I_{ma}^{1}, \circ, I_{pl}^{1}>, <I_{ph}^{1}, \circ, I_{pl}^{1}>\},$
$\{<I_{ph}^{1}, \prec^{\circ}, I_{ph}^{9}>, <I_{ma}^{1}, \prec^{\circ}, I_{ma}^{9}>, <I_{pl}^{1}, \prec^{\circ}, I_{pl}^{9}>\},$
$9>$

$<1,$
$\{<I_{ph}^{1}, <\text{PHONATION:voiceless}>>, <I_{ma}^{1}, <\text{MANNER:plosive}>>\},$
$\{<I_{ph}^{1}, \circ, I_{ma}^{1}>\},$
$\{<I_{ma}^{1}, \prec^{\circ}, I_{ma}^{peak}>\},$
$\text{peak}>$

$<1,$
$\{<I_{ph}^{1}, <\text{PHONATION:voiced}>>, <I_{ma}^{1}, <\text{MANNER:nasal}>>, <I_{pl}^{1}, <\text{PLACE: } \alpha>>\} \mid \alpha \in \{\text{labial, apical}\},$
$\{<I_{ph}^{1}, \circ, I_{ma}^{1}>, <I_{ma}^{1}, \circ, I_{pl}^{1}>, <I_{ph}^{1}, \circ, I_{pl}^{1}>\},$
$\{<I_{ma}^{1}, \prec^{\circ}, I_{ma}^{peak}>, <I_{pl}^{1}, \prec^{\circ}, I_{pl}^{peak}>\},$
$\text{peak}>$

$<1,$
$\{<I_{ph}^{1}, <\text{PHONATION:voiced}>>, <I_{ma}^{1}, <\text{MANNER:fricative}>>, <I_{pl}^{1}, <\text{PLACE: labial}>>\},$
$\{I_{ph}^{1}, \circ, I_{ma}^{1}>, <I_{ma}^{1}, \circ, I_{pl}^{1}>, <I_{ph}^{1}, \circ, I_{pl}^{1}>\},$
$\{<I_{ph}^{1}, \prec^{\circ}, I_{ph}^{peak}>, <I_{ma}^{1}, \prec^{\circ}, I_{ma}^{peak}>, <I_{pl}^{1}, \prec^{\circ}, I_{pl}^{peak}>\},$
$\text{peak}>$

$<2,$
$\{<I_{ph}^{2}, <\text{PHONATION:voiceless}>>, <I_{ma}^{2}, <\text{MANNER:plosive}>>, <I_{pl}^{2}, <\text{PLACE: velar}>>\},$
$\{<I_{ph}^{2}, \circ, I_{ma}^{2}>, <I_{ma}^{2}, \circ, I_{pl}^{2}>, <I_{ph}^{2}, \circ, I_{pl}^{2}>\},$
$\{<I_{ph}^{2}, \prec^{\circ}, I_{ph}^{10}>, <I_{ma}^{2}, \prec^{\circ}, I_{ma}^{10}>, <I_{pl}^{2}, \prec^{\circ}, I_{pl}^{10}>\},$

10>

<10,
{<I_{ph}^{10}, <PHONATION:voiced>>, <I_{ma}^{10}, <MANNER:lateral>>},
{<I_{ph}^{10}, $\circ$,I_{ma}^{10}>},
{<I_{ph}^{10},$\prec\circ$,I_{ph}^{peak}>, <I_{ma}^{10},$\prec\circ$,I_{ma}^{peak}>},
peak>

<10,
{<I_{ph}^{10}, <PHONATION:voiced>>, <I_{ma}^{10}, <MANNER:fricative>>, <I_{pl}^{10}, <PLACE: uvular>>},
{<I_{ph}^{10}, $\circ$,I_{ma}^{10}>, <I_{ma}^{10}, $\circ$,I_{pl}^{10}>, <I_{ph}^{10}, $\circ$,I_{pl}^{10}>},
{<I_{ph}^{10},$\prec\circ$,I_{ph}^{peak}>, <I_{ma}^{10},$\prec\circ$,I_{ma}^{peak}>, <I_{pl}^{10},$\prec\circ$,I_{pl}^{peak}>},
peak>

<2,
{<I_{ph}^{2}, <PHONATION:voiceless>>, <I_{ma}^{2}, <MANNER:fricative>>, <I_{pl}^{2}, <PLACE: labial>>},
{<I_{ph}^{2}, $\circ$,I_{ma}^{2}>, <I_{ma}^{2}, $\circ$,I_{pl}^{2}>, <I_{ph}^{2}, $\circ$,I_{pl}^{2}>},
{<I_{ph}^{2},$\prec\circ$,I_{ph}^{peak}>, <I_{ma}^{2},$\prec\circ$,I_{ma}^{peak}>, <I_{pl}^{2},$\prec\circ$,I_{pl}^{peak}>},
peak>

<2,
{<I_{ph}^{2}, <PHONATION:voiceless>>, <I_{ma}^{2}, <MANNER:affricate>>, <I_{pl}^{2}, <PLACE: apical>>},
{<I_{ph}^{2}, $\circ$,I_{ma}^{2}>, <I_{ma}^{2}, $\circ$,I_{pl}^{2}>, <I_{ph}^{2}, $\circ$,I_{pl}^{2}>},
{<I_{ph}^{2},$\prec\circ$,I_{ph}^{peak}>, <I_{ma}^{2},$\prec\circ$,I_{ma}^{peak}>, <I_{pl}^{2},$\prec\circ$,I_{pl}^{peak}>},
peak>

<2,
{<I_{ph}^{2}, <PHONATION:voiced>>, <I_{ma}^{2}, <MANNER:lateral>>},
{<I_{ph}^{2}, $\circ$,I_{ma}^{2}>},
{<I_{ph}^{2},$\prec\circ$,I_{ph}^{peak}>, <I_{ma}^{2},$\prec\circ$,I_{ma}^{peak}>},
peak>

<3,
{<I_{ph}^{3}, <PHONATION:voiced>>, <I_{ma}^{3}, <MANNER:fricative>>, <I_{pl}^{3}, <PLACE: labial>>},
{<I_{ph}^{3}, $\circ$,I_{ma}^{3}>, <I_{ma}^{3}, $\circ$,I_{pl}^{3}>, <I_{ph}^{3}, $\circ$,I_{pl}^{3}>},
{<I_{ma}^{3},$\prec\circ$,I_{ma}^{peak}>, <I_{pl}^{3},$\prec\circ$,I_{pl}^{peak}>},
peak>

<4,
{<I_{ph}^{4}, <PHONATION:voiced>>, <I_{ma}^{4}, <MANNER:lateral>>},
{<I_{ph}^{4}, $\circ$,I_{ma}^{4}>},
{<I_{ma}^{4},$\prec\circ$,I_{ma}^{peak}>},
peak>

<5,
{<I_{ph}^{5}, <PHONATION:voiced>>, <I_{ma}^{5}, <MANNER:fricative>>, <I_{pl}^{5}, <PLACE: uvular>>},
{<I_{ph}^{5}, $\circ$,I_{ma}^{5}>, <I_{ma}^{5}, $\circ$,I_{pl}^{5}>, <I_{ph}^{5}, $\circ$,I_{pl}^{5}>},
{<I_{ma}^{5},$\prec\circ$,I_{ma}^{peak}>, <I_{pl}^{5},$\prec\circ$,I_{pl}^{peak}>},
peak>

<6,
{<I_{ph}^{6}, <PHONATION:voiced>>, <I_{ma}^{6}, <MANNER:nasal>>, <I_{pl}^{6}, <PLACE: apical>>},
{<I_{ph}^{6}, $\circ$,I_{ma}^{6}>, <I_{ma}^{6}, $\circ$,I_{pl}^{6}>, <I_{ph}^{6}, $\circ$,I_{pl}^{6}>},

$\{<I_{ma}^{6}, \prec^{\circ}, I_{ma}^{peak}>, <I_{pl}^{6}, \prec^{\circ}, I_{pl}^{peak}>\},$
$\text{peak}>$

$<7,$
$\{<I_{ph}^{7}, <\text{PHONATION:voiced}>>, <I_{ma}^{7}, <\text{MANNER:nasal}>>, <I_{pl}^{7}, <\text{PLACE: apical}>>\},$
$\{<I_{ph}^{7}, ^{\circ}, I_{ma}^{7}>, <I_{ma}^{7}, ^{\circ}, I_{pl}^{7}>, <I_{ph}^{7}, ^{\circ}, I_{pl}^{7}>\},$
$\{<I_{ma}^{7}, \prec^{\circ}, I_{ma}^{peak}>, <I_{pl}^{7}, \prec^{\circ}, I_{pl}^{peak}>\},$
$\text{peak}>$

$<7,$
$\{<I_{ph}^{7}, <\text{PHONATION:voiceless}>>, <I_{ma}^{7}, <\text{MANNER:fricative}>>, <I_{pl}^{7}, <\text{PLACE: apical}>>\},$
$\{<I_{ph}^{7}, ^{\circ}, I_{ma}^{7}>, <I_{ma}^{7}, ^{\circ}, I_{pl}^{7}>, <I_{ph}^{7}, ^{\circ}, I_{pl}^{7}>\},$
$\{<I_{ma}^{7}, \prec^{\circ}, I_{ma}^{peak}>, <I_{pl}^{7}, \prec^{\circ}, I_{pl}^{peak}>\},$
$\text{peak}>$

$<8,$
$\{<I_{ph}^{8}, <\text{PHONATION:voiced}>>, <I_{ma}^{8}, <\text{MANNER:lateral}>>\},$
$\{<I_{ph}^{8}, ^{\circ}, I_{ma}^{8}>\},$
$\{<I_{ma}^{8}, \prec^{\circ}, I_{ma}^{peak}>\},$
$\text{peak}>$

$<8,$
$\{<I_{ph}^{8}, <\text{PHONATION:voiced}>>, <I_{ma}^{8}, <\text{MANNER:fricative}>>, <I_{pl}^{8}, <\text{PLACE: uvular}>>\},$
$\{<I_{ph}^{8}, ^{\circ}, I_{ma}^{8}>, <I_{ma}^{8}, ^{\circ}, I_{pl}^{8}>, <I_{ph}^{8}, ^{\circ}, I_{pl}^{8}>\},$
$\{<I_{ma}^{8}, \prec^{\circ}, I_{ma}^{peak}>, <I_{pl}^{8}, \prec^{\circ}, I_{pl}^{peak}>\},$
$\text{peak}>$

$<9,$
$\{<I_{ph}^{9}, <\text{PHONATION:voiced}>>, <I_{ma}^{9}, <\text{MANNER:fricative}>>, <I_{pl}^{9}, <\text{PLACE: uvular}>>\},$
$\{<I_{ph}^{9}, ^{\circ}, I_{ma}^{9}>, <I_{ma}^{9}, ^{\circ}, I_{pl}^{9}>, <I_{ph}^{9}, ^{\circ}, I_{pl}^{9}>\},$
$\{<I_{ma}^{9}, \prec^{\circ}, I_{ma}^{peak}>, <I_{pl}^{9}, \prec^{\circ}, I_{pl}^{peak}>\},$
$\text{peak}>$

$<\text{peak},$
$\{<I_{ma}^{peak}, <\text{MANNER:vowellike}>>, <I_{le}^{peak}, <\text{LENGTH:long}>>\},$
$\{<I_{ma}^{peak}, ^{\circ}, I_{le}^{peak}>\},$
$\{\},$
$\text{end}>$

$<\text{peak},$
$\{<I_{ma}^{peak}, <\text{MANNER:diphthong}>>\},$
$\{\},$
$\{\},$
$\text{end}>$

$<\text{peak},$
$\{<I_{ma}^{peak}, <\text{MANNER:vowellike}>>, <I_{pl}^{peak}, <\text{PLACE: } \alpha>>\} \mid \alpha \in \{\text{front, back}\},$
$\{<I_{ma}^{peak}, ^{\circ}, I_{pl}^{peak}>\},$
$\{<I_{ma}^{peak}, \prec^{\circ}, I_{ma}^{11}>, <I_{pl}^{peak}, \prec^{\circ}, I_{pl}^{11}>\},$
$11>$

$<\text{peak},$
$\{<<I_{ma}^{peak}, <\text{MANNER:diphthong}>>\},$
$\{\},$
$\{<I_{ma}^{peak}, \prec^{\circ}, I_{ma}^{11}>\},$

11>

<peak,
{<I_{ma}^{peak}, <MANNER:vowellike>>, <I_{pl}^{peak}, <PLACE: front>>},
{<I_{ma}^{peak}, $\circ$, I_{pl}^{peak}>},
{<I_{ma}^{peak}, $\prec^{\circ}$, I_{ma}^{12}>, <I_{pl}^{peak}, $\prec^{\circ}$, I_{pl}^{12}>},
12>

<peak,
{<I_{ma}^{peak}, <MANNER:vowellike>>, <I_{pl}^{peak}, <PLACE: back>>},
{<I_{ma}^{peak}, $\circ$, I_{pl}^{peak}>},
{<I_{ma}^{peak}, $\prec^{\circ}$, I_{ma}^{13}>, <I_{pl}^{peak}, $\prec^{\circ}$, I_{pl}^{13}>},
13>

<peak,
{<I_{ma}^{peak}, <MANNER:vowellike>>, <I_{pl}^{peak}, <PLACE: α>} | $\alpha \in$ {front, back>>,
{<I_{ma}^{peak}, $\circ$, I_{pl}^{peak}>},
{<I_{ma}^{peak}, $\prec^{\circ}$, I_{ma}^{14}>, <I_{pl}^{peak}, $\prec^{\circ}$, I_{pl}^{14}>},
14>

<peak,
{<I_{ma}^{peak}, <MANNER:diphthong>>},
{},
{<I_{ma}^{peak}, $\circ$, I_{ma}^{14}>},
14>

<11,
{<I_{ph}^{11}, <PHONATION:voiceless>>, <I_{ma}^{11}, <MANNER:affricate>>,, <I_{pl}^{11}, <PLACE: α>>} | $\alpha \in$ {apical, palato-alveolar},
{<I_{ph}^{11}, $\circ$, I_{ma}^{11}>, <I_{ma}^{11}, $\circ$, I_{pl}^{11}>, <I_{ph}^{11}, $\circ$, I_{pl}^{11}>},
{},
end>

<12,
{<I_{ph}^{12}, <PHONATION:voiceless>>, <I_{ma}^{12}, <MANNER:fricative>>, <I_{pl}^{12}, <PLACE: palatal>>},
{<I_{ph}^{12}, $\circ$, I_{ma}^{12}>, <I_{ma}^{12}, $\circ$, I_{pl}^{12}>, <I_{ph}^{12}, $\circ$, I_{pl}^{12}>},
{<I_{ph}^{12}, $\prec^{\circ}$, I_{ph}^{22}>, <I_{ma}^{12}, $\prec^{\circ}$, I_{ma}^{22}>, <I_{pl}^{12}, $\prec^{\circ}$, I_{pl}^{22}>},
22>

<13,
{<I_{ph}^{13}, <PHONATION:voiceless>>, <I_{ma}^{13}, <MANNER:fricative>>, <I_{pl}^{13}, <PLACE: velar>>},
{<I_{ph}^{13}, $\circ$, I_{ma}^{13}>, <I_{ma}^{13}, $\circ$, I_{pl}^{13}>, <I_{ph}^{13}, $\circ$, I_{pl}^{13}>},
{<I_{ph}^{13}, $\prec^{\circ}$, I_{ph}^{22}>, <I_{ma}^{13}, $\prec^{\circ}$, I_{ma}^{22}>, <I_{pl}^{13}, $\prec^{\circ}$, I_{pl}^{22}>},
22>

<14,
{<I_{ph}^{14}, <PHONATION:voiced>>, <I_{ma}^{14}, <MANNER nasal>>, <I_{pl}^{14}, <PLACE: α>>} | $\alpha \in$ {velar, apical, labial}
{<I_{ph}^{14}, $\circ$, I_{ma}^{14}>, <I_{ma}^{14}, $\circ$, I_{pl}^{14}>, <I_{ph}^{14}, $\circ$, I_{pl}^{14}>},
{<I_{ph}^{14}, $\prec^{\circ}$, I_{ph}^{15}>, <I_{ma}^{14}, $\prec^{\circ}$, I_{ma}^{15}>, <I_{pl}^{14}, $\prec^{\circ}$ I_{pl}^{15}},
15>

<14,
{<I_{ph}^{14}, <PHONATION:voiced>>, <I_{ma}^{14}, <MANNER:nasal>>, <I_{pl}^{14}, <PLACE: apical>>},

$\{<I_{ph}^{14}, \circ, I_{ma}^{14}>, <I_{ma}^{14}, \circ, I_{pl}^{14}>, <I_{ph}^{14}, \circ, I_{pl}^{14}>\},$
$\{<I_{ph}^{14}, \prec\circ, I_{ph}^{16}>, <I_{ma}^{14}, \prec\circ, I_{ma}^{16}>, <I_{pl}^{14}, \prec\circ, I_{pl}^{16}>\},$
$16>$

$<14,$
$\{<I_{ph}^{14}, <\text{PHONATION:voiced}>>, <I_{ma}^{14}, <\text{MANNER:nasal}>>, <I_{pl}^{14}, <\text{PLACE: labial}>>\},$
$\{<I_{ph}^{14}, \circ, I_{ma}^{14}>, <I_{ma}^{14}, \circ, I_{pl}^{14}>, <I_{ph}^{14}, \circ, I_{pl}^{14}>\},$
$\{<I_{ph}^{14}, \prec\circ, I_{ph}^{17}>, <I_{ma}^{14}, \prec\circ, I_{ma}^{17}>, <I_{pl}^{14}, \prec\circ, I_{pl}^{17}>\},$
$17>$

$<14,$
$\{<I_{ph}^{14}, <\text{PHONATION:voiceless}>>, <I_{ma}^{14}, <\text{MANNER:plosive}>>\},$
$\{<I_{ph}^{14}, \circ, I_{ma}^{14}>\},$
$\{<I_{ph}^{14}, \prec\circ, I_{ph}^{18}>, <I_{ma}^{14}, \prec\circ, I_{ma}^{18}>\},$
$18>$

$<14,$
$\{<I_{ph}^{14}, <\text{PHONATION:voiced}>>, <I_{ma}^{14}, <\text{MANNER:nasal}>>, <I_{pl}^{14}, <\text{PLACE: } \alpha>>\}$
$\mid \alpha \in \{\text{labial, apical}\},$
$\{<I_{ph}^{14}, \circ, I_{ma}^{14}>, <I_{ma}^{14}, \circ, I_{pl}^{14}>, <I_{ph}^{14}, \circ, I_{pl}^{14}>\},$
$\{<I_{ph}^{14}, \prec\circ, I_{ph}^{18}>, <I_{ma}^{14}, \prec\circ, I_{ma}^{18}>, <I_{pl}^{14}, \prec\circ, I_{pl}^{18}>\},$
$18>$

$<14,$
$\{<I_{ph}^{14}, <\text{PHONATION:voiced}>>, <I_{ma}^{14}, <\text{MANNER:lateral}>>\},$
$\{<I_{ph}^{14}, \circ, I_{ma}^{14}>\},$
$\{<I_{ph}^{14}, \prec\circ, I_{ph}^{18}>, <I_{ma}^{14}, \prec\circ, I_{ma}^{18}>\},$
$18>$

$<14,$
$\{<I_{ph}^{14}, <\text{PHONATION:voiced}>>, <I_{ma}^{14}, <\text{MANNER:fricative}>>, <I_{pl}^{14}, <\text{PLACE: uvular}>>\},$
$\{<I_{ph}^{14}, \circ, I_{ma}^{14}>, <I_{ma}^{14}, \circ, I_{pl}^{14}>, <I_{ph}^{14}, \circ, I_{pl}^{14}>\},$
$\{<I_{ph}^{14}, \prec\circ, I_{ph}^{18}>, <I_{ma}^{14}, \prec\circ, I_{ma}^{18}>, <I_{pl}^{14}, \prec\circ, I_{pl}^{18}>\},$
$18>$

$<14,$
$\{<I_{ph}^{14}, <\text{PHONATION:voiced}>>, <I_{ma}^{14}, <\text{MANNER:lateral}>>\},$
$\{<I_{ph}^{14}, \circ, I_{ma}^{14}>\},$
$\{<I_{ph}^{14}, \prec\circ, I_{ph}^{19}>, <I_{ma}^{14}, \prec\circ, I_{ma}^{19}>\},$
$19>$

$<14,$
$\{<I_{ph}^{14}, <\text{PHONATION:voiced}>>, <I_{ma}^{14}, <\text{MANNER:fricative}>>, <I_{pl}^{14}, <\text{PLACE: uvular}>>\},$
$\{<I_{ph}^{14}, \circ, I_{ma}^{14}>, <I_{ma}^{14}, \circ, I_{pl}^{14}>, <I_{ph}^{14}, \circ, I_{pl}^{14}>\},$
$\{<I_{ph}^{14}, \prec\circ, I_{ph}^{19}>, <I_{ma}^{14}, \prec\circ, I_{ma}^{19}>, <I_{pl}^{14}, \prec\circ, I_{pl}^{19}>\},$
$19>$

$<14,$
$\{<I_{ph}^{14}, <\text{PHONATION:voiced}>>, <I_{ma}^{14}, <\text{MANNER:fricative}>>, <I_{pl}^{14}, <\text{PLACE: uvular}>>\},$
$\{<I_{ph}^{14}, \circ, I_{ma}^{14}>, <I_{ma}^{14}, \circ, I_{pl}^{14}>, <I_{ph}^{14}, \circ, I_{pl}^{14}>\},$
$\{<I_{ph}^{14}, \prec\circ, I_{ph}^{20}>, <I_{ma}^{14}, \prec\circ, I_{ma}^{20}>, <I_{pl}^{14}, \prec\circ, I_{pl}^{20}>\},$
$20>$

<14,
{<I_{ph}^{14}, <PHONATION:voiceless>>, <I_{ma}^{14}, <MANNER:fricative>>, <I_{pl}^{14}, <PLACE: apical>>},
{<I_{ph}^{14},$^\circ$,I_{ma}^{14}>, <I_{ma}^{14},$^\circ$,I_{pl}^{14}>, <I_{ph}^{14},$^\circ$,I_{pl}^{14}>},
{<I_{ph}^{14},$\prec^\circ$,I_{ph}^{21}>, <I_{ma}^{14},$\prec^\circ$,I_{ma}^{21}>, <I_{pl}^{14},$\prec^\circ$,I_{pl}^{21}>},
21>

<14,
{<I_{ph}^{14}, <PHONATION:voiceless>>, <I_{ma}^{14}, <MANNER:plosive>>, <I_{pl}^{14}, <PLACE: α>>} | $\alpha \in$ {labial, velar},
{<I_{ph}^{14},$^\circ$,I_{ma}^{14}>, <I_{ma}^{14},$^\circ$,I_{pl}^{14}>, <I_{ph}^{14},$^\circ$,I_{pl}^{14}>},
{<I_{ph}^{14},$\prec^\circ$,I_{ph}^{22}>, <I_{ma}^{14},$\prec^\circ$,I_{ma}^{22}7>, <I_{pl}^{14},$\prec^\circ$,I_{pl}^{22}>},
22>

<14,
{<I_{ph}^{14}, <PHONATION:voiceless>>, <I_{ma}^{14}, <MANNER:fricative>>, <I_{pl}^{14}, <PLACE: α>>} | $\alpha \in$ {labial, palato},
{<I_{ph}^{14},$^\circ$,I_{ma}^{14}>, <I_{ma}^{14},$^\circ$,I_{pl}^{14}>, <I_{ph}^{14},$^\circ$,I_{pl}^{14}>},
{<I_{ph}^{14},$\prec^\circ$,I_{ph}^{22}>, <I_{ma}^{14},$\prec^\circ$,I_{ma}^{22}>, <I_{pl}^{14},$\prec^\circ$,I_{pl}^{22}>},
22>

<14,
{<I_{ph}^{14}, <PHONATION:voiced>>, <I_{ma}^{14}, <MANNER:nasal>>},
{<I_{ph}^{14},$^\circ$,I_{ma}^{14}>},
{<I_{ph}^{14},$\prec^\circ$,I_{ph}^{22}>, <I_{ma}^{14},$\prec^\circ$,I_{ma}^{22}>},
22>

<14,
{<I_{ph}^{14}, <PHONATION:voiceless>>, <I_{ma}^{14}, <MANNER:affricate>>, <I_{pl}^{14}, <PLACE: labial>>},
{<I_{ph}^{14},$^\circ$,I_{ma}^{14}>, <I_{ma}^{14},$^\circ$,I_{pl}^{14}>, <I_{ph}^{14},$^\circ$,I_{pl}^{14}>},
{<I_{ph}^{14},$\prec^\circ$,I_{ph}^{22}>, <I_{ma}^{14},$\prec^\circ$,I_{ma}^{22}>, <I_{pl}^{14},$\prec^\circ$,I_{pl}^{22}>},
22>

<14,
{<I_{ph}^{14}, <PHONATION:voiced>>, <I_{ma}^{14}, <MANNER:lateral>>},
{<I_{ph}^{14},$^\circ$,I_{ma}^{14}>},
{<I_{ph}^{14},$\prec^\circ$,I_{ph}^{22}>, <I_{ma}^{14},$\prec^\circ$,I_{ma}^{22}>},
22>

<14,
{<I_{ph}^{14}, <PHONATION:voiced>>, <I_{ma}^{14}, <MANNER:fricative>>, <I_{pl}^{14}, <PLACE: uvular>>},
{<I_{ph}^{14},$^\circ$,I_{ma}^{14}>, <I_{ma}^{14},$^\circ$,I_{pl}^{14}>, <I_{ph}^{14},$^\circ$,I_{pl}^{14}>},
{<I_{ph}^{14},$\prec^\circ$,I_{ph}^{22}>, <I_{ma}^{14},$\prec^\circ$,I_{ma}^{22}>, <I_{pl}^{14},$\prec^\circ$,I_{pl}^{22}>},
22>

<14,
{<I_{ph}^{14}, <PHONATION:voiceless>>, <I_{ma}^{14}, <MANNER:plosive>>, <I_{pl}^{14}, <PLACE: apical>>},
{<I_{ph}^{14},$^\circ$,I_{ma}^{14}>, <I_{ma}^{14},$^\circ$,I_{pl}^{14}>, <I_{ph}^{14},$^\circ$,I_{pl}^{14}>},
{<I_{ph}^{14},$\prec^\circ$,I_{ph}^{22}>, <I_{ma}^{14},$\prec^\circ$,I_{ma}^{22}>, <I_{pl}^{14},$\prec^\circ$,I_{pl}^{22}>},
22>

<14,
{<I_{ph}^{14}, <PHONATION:voiceless>>, <I_{ma}^{14}, <MANNER:fricative>>, <I_{pl}^{14}, <PLACE: apical>>},

{<I_{ph}^{14},$\circ$,I_{ma}^{14}>, <I_{ma}^{14},$\circ$,I_{pl}^{14}>, <I_{ph}^{14},$\circ$,I_{pl}^{14}>},
{<I_{ph}^{14},$\prec\circ$,I_{ph}^{22}>, <I_{ma}^{14},$\prec\circ$,I_{ma}^{22}>, <I_{pl}^{14},$\prec\circ$,I_{pl}^{22}>},
22>

<14,
{<I_{ph}^{14}, <PHONATION:voiced>>, <I_{ma}^{14}, <MANNER:nasal>>},
{<I_{ph}^{14},$\circ$,I_{ma}^{14}>},
{<I_{ph}^{14},$\prec\circ$,I_{ph}^{23}>, <I_{ma}^{14},$\prec\circ$,I_{ma}^{23}>},
23>

<14,
{<I_{ph}^{14}, <PHONATION:voiceless>>, <I_{ma}^{14}, <MANNER:plosive>>, <I_{pl}^{14}, <PLACE: labial>>},
{<I_{ph}^{14},$\circ$,I_{ma}^{14}>, <I_{ma}^{14},$\circ$,I_{pl}^{14}>, <I_{ph}^{14},$\circ$,I_{pl}^{14}>},
{<I_{ph}^{14},$\prec\circ$,I_{ph}^{23}>, <I_{ma}^{14},$\prec\circ$,I_{ma}^{23}>, <I_{pl}^{14},$\prec\circ$,I_{pl}^{23}>},
23>

<14,
{<I_{ph}^{14}, <PHONATION:voiced>>, <I_{ma}^{14}, <MANNER:lateral>>>>},
{<I_{ph}^{14},$\circ$,I_{ma}^{14}>},
{<I_{ph}^{14},$\prec\circ$,I_{ph}^{23}>, <I_{ma}^{14},$\prec\circ$,I_{ma}^{23}>},
23>

<14,
{<I_{ph}^{14}, <PHONATION:voiced>>, <I_{ma}^{14}, <MANNER:fricative>>, <I_{pl}^{14}, <PLACE: uvular>>},
{<I_{ph}^{14},$\circ$,I_{ma}^{14}>, <I_{ma}^{14},$\circ$,I_{pl}^{14}>, <I_{ph}^{14},$\circ$,I_{pl}^{14}>},
{<I_{ph}^{14},$\prec\circ$,I_{ph}^{23}>, <I_{ma}^{14},$\prec\circ$,I_{ma}^{23}>, <I_{pl}^{14},$\prec\circ$,I_{pl}^{23}>},
23>

<15,
{<I_{ph}^{15}, <PHONATION:voiceless>>, <I_{ma}^{15}, <MANNER:plosive>>, <I_{pl}^{15}, <PLACE: apical>>},
{<I_{ph}^{15},$\circ$,I_{ma}^{15}>, <I_{ma}^{15},$\circ$,I_{pl}^{15}>, <I_{ph}^{15},$\circ$,I_{pl}^{15}>},
{<I_{ph}^{15},$\prec\circ$,I_{ph}^{22}>, <I_{ma}^{15},$\prec\circ$,I_{ma}^{22}>, <I_{pl}^{15},$\prec\circ$,I_{pl}^{22}>},
22>

<15,
{<I_{ph}^{15}, <PHONATION:voiceless>>, <I_{ma}^{15}, <MANNER:plosive>>, <I_{pl}^{15}, <PLACE: α>>} | $\alpha \in$ {labial, velar},
{<I_{ph}^{15},$\circ$,I_{ma}^{15}>, <I_{ma}^{15},$\circ$,I_{pl}^{15}>, <I_{ph}^{15},$\circ$,I_{pl}^{15}>},
{<I_{ph}^{15},$\prec\circ$,I_{ph}^{22}>, <I_{ma}^{15},$\prec\circ$,I_{ma}^{22}>, <I_{pl}^{15},$\prec\circ$,I_{pl}^{22}>},
22>

<16,
{<I_{ph}^{16}, <PHONATION:voiceless>>, <I_{ma}^{16}, <MANNER:affricate>>, <I_{pl}^{16}, <PLACE: palato-alveolar>>},
{<I_{ph}^{16},$\circ$,I_{ma}^{16}>, <I_{ma}^{16},$\circ$,I_{pl}^{16}>, <I_{ph}^{16},$\circ$,I_{pl}^{16}>},
{<I_{ph}^{16},$\prec\circ$,I_{ph}^{22}>, <I_{ma}^{16},$\prec\circ$,I_{ma}^{22}>, <I_{pl}^{16},$\prec\circ$,I_{pl}^{22}>},
22>

<16,
{<I_{ph}^{16}, <PHONATION:voiceless>>, <I_{ma}^{16}, <MANNER:fricative>>, <I_{pl}^{16}, <PLACE: α>>} | $\alpha \in$ {labial, palato-alveolar},
{<I_{ph}^{16},$\circ$,I_{ma}^{16}>, <I_{ma}^{16},$\circ$,I_{pl}^{16}>, <I_{ph}^{16},$\circ$,I_{pl}^{16}>},
{<I_{ph}^{16},$\prec\circ$,I_{ph}^{22}>, <I_{ma}^{16},$\prec\circ$,I_{ma}^{22}>, <I_{pl}^{16},$\prec\circ$,I_{pl}^{22}>},
22>

<17,
{<$I_{ph}{}^{17}$, <PHONATION:voiceless>>, <$I_{pl}{}^{17}$, <PLACE: labial>>, <$I_{ma}{}^{17}$, <MANNER: affricate>>},
{<$I_{ph}{}^{17}$,$\circ$,$I_{ma}{}^{17}$>, <$I_{ma}{}^{17}$,$\circ$,$I_{pl}{}^{17}$>, <$I_{ph}{}^{17}$,$\circ$,$I_{pl}{}^{17}$>},
{<$I_{ph}{}^{17}$,$\prec^{\circ}$,$I_{ph}{}^{22}$>, <$I_{ma}{}^{17}$,$\prec^{\circ}$,$I_{ma}{}^{22}$>, <$I_{pl}{}^{17}$,$\prec^{\circ}$,$I_{pl}{}^{22}$>},
22>

<18,
{<$I_{ph}{}^{18}$, <PHONATION:voiceless>>, <$I_{ma}{}^{18}$, <MANNER:fricative>>, <$I_{pl}{}^{18}$, <PLACE: palato-alveolar>>},
{<$I_{ph}{}^{18}$,$\circ$,$I_{ma}{}^{18}$>, <$I_{ma}{}^{18}$,$\circ$,$I_{pl}{}^{18}$>, <$I_{ph}{}^{18}$,$\circ$,$I_{pl}{}^{18}$>},
{<$I_{ph}{}^{18}$,$\prec^{\circ}$,$I_{ph}{}^{22}$>, <$I_{ma}{}^{18}$,$\prec^{\circ}$,$I_{ma}{}^{22}$>, <$I_{pl}{}^{18}$,$\prec^{\circ}$,$I_{pl}{}^{22}$>},
22>

<19,
{<$I_{ph}{}^{19}$, <PHONATION:voiceless>>, <$I_{ma}{}^{19}$, <MANNER:plosive>>, <$I_{pl}{}^{19}$, <PLACE: α>>} | $\alpha \in$ {labial, velar},
{<$I_{ph}{}^{19}$,$\circ$,$I_{ma}{}^{19}$>, <$I_{ma}{}^{19}$,$\circ$,$I_{pl}{}^{19}$>, <$I_{ph}{}^{19}$,$\circ$,$I_{pl}{}^{19}$>},
{<$I_{ph}{}^{19}$,$\prec^{\circ}$,$I_{ph}{}^{22}$>, <$I_{ma}{}^{19}$,$\prec^{\circ}$,$I_{ma}{}^{22}$>, <$I_{pl}{}^{19}$,$\prec^{\circ}$,$I_{pl}{}^{22}$>},
22>

<19,
{<$I_{ph}{}^{19}$, <PHONATION:voiced>>, <$I_{ma}{}^{19}$, <MANNER:nasal>>, <$I_{pl}{}^{19}$, <PLACE: α>>} | $\alpha \in$ {labial, apical},
{<$I_{ph}{}^{19}$,$\circ$,$I_{ma}{}^{19}$>, <$I_{ma}{}^{19}$,$\circ$,$I_{pl}{}^{19}$>, <$I_{ph}{}^{19}$,$\circ$,$I_{pl}{}^{19}$>},
{<$I_{ph}{}^{19}$,$\prec^{\circ}$,$I_{ph}{}^{22}$>, <$I_{ma}{}^{19}$,$\prec^{\circ}$,$I_{ma}{}^{22}$>, <$I_{pl}{}^{19}$,$\prec^{\circ}$,$I_{pl}{}^{22}$>},
22>

<19,
{<$I_{ph}{}^{19}$, <PHONATION:voiceless>>, <$I_{ma}{}^{19}$, <MANNER:fricative>>, <$I_{pl}{}^{19}$, <PLACE: α>>} | $\alpha \in$ {labial, palatal},
{<$I_{ph}{}^{19}$,$\circ$,$I_{ma}{}^{19}$>, <$I_{ma}{}^{19}$,$\circ$,$I_{pl}{}^{19}$>, <$I_{ph}{}^{19}$,$\circ$,$I_{pl}{}^{19}$>},
{<$I_{ph}{}^{19}$,$\prec^{\circ}$,$I_{ph}{}^{22}$>, <$I_{ma}{}^{19}$,$\prec^{\circ}$,$I_{ma}{}^{22}$>, <$I_{pl}{}^{19}$,$\prec^{\circ}$,$I_{pl}{}^{22}$>},
22>

<19,
{<$I_{ph}{}^{19}$, <PHONATION:voiceless>>, <$I_{ma}{}^{19}$, <MANNER:affricate>>, <$I_{pl}{}^{19}$, <PLACE: apical>>},
{<$I_{ph}{}^{19}$,$\circ$,$I_{ma}{}^{19}$>, <$I_{ma}{}^{19}$,$\circ$,$I_{pl}{}^{19}$>, <$I_{ph}{}^{19}$,$\circ$,$I_{pl}{}^{19}$>},
{<$I_{ph}{}^{19}$,$\prec^{\circ}$,$I_{ph}{}^{22}$>, <$I_{ma}{}^{19}$,$\prec^{\circ}$,$I_{ma}{}^{22}$>, <$I_{pl}{}^{19}$,$\prec^{\circ}$,$I_{pl}{}^{22}$>},
22>

<20,
{<$I_{ph}{}^{20}$, <PHONATION:voiced>>, <$I_{ma}{}^{20}$, <MANNER:lateral>>>>},
{<$I_{ph}{}^{20}$,$\circ$,$I_{ma}{}^{20}$>},
{<$I_{ph}{}^{20}$,$\prec^{\circ}$,$I_{ph}{}^{24}$>, <$I_{ma}{}^{20}$,$\prec^{\circ}$,$I_{ma}{}^{24}$>},
24>

<20,
{<$I_{ph}{}^{20}$, <PHONATION:voiceless>>, <$I_{ma}{}^{20}$, <MANNER:fricative>>, <$I_{pl}{}^{20}$, <PLACE: α>>} | $\alpha \in$ {labial, palatal},
{<$I_{ph}{}^{20}$,$\circ$,$I_{ma}{}^{20}$>, <$I_{ma}{}^{20}$,$\circ$,$I_{pl}{}^{20}$>, <$I_{ph}{}^{20}$,$\circ$,$I_{pl}{}^{20}$>},
{<$I_{ph}{}^{20}$,$\prec^{\circ}$,$I_{ph}{}^{23}$>, <$I_{ma}{}^{20}$,$\prec^{\circ}$,$I_{ma}{}^{23}$>, <$I_{pl}{}^{20}$,$\prec^{\circ}$,$I_{pl}{}^{23}$>},
23>

<20,
{<$I_{ph}{}^{20}$, <PHONATION:voiceless>>, <$I_{ma}{}^{20}$, <MANNER:plosive>>, <$I_{pl}{}^{20}$, <PLACE: labial>>},
{<$I_{ph}{}^{20}$,$\circ$,$I_{ma}{}^{20}$>, <$I_{ma}{}^{20}$,$\circ$,$I_{pl}{}^{20}$>, <$I_{ph}{}^{20}$,$\circ$,$I_{pl}{}^{20}$>},

{<I_{ph}^{20},$\prec^{\circ}$,I_{ph}^{23}>, <I_{ma}^{20},$\prec^{\circ}$,I_{ma}^{23}>, <I_{pl}^{20},$\prec^{\circ}$,I_{pl}^{23}>},
23>

<20,
{<I_{ph}^{20}, <PHONATION:voiced>>, <I_{ma}^{20}, <MANNER:nasal>>, <I_{pl}^{20}, <PLACE: apical>>},
{<I_{ph}^{20},$\circ$,I_{ma}^{20}>, <I_{ma}^{20},$\circ$,I_{pl}^{20}>, <I_{ph}^{20},$\circ$,I_{pl}^{20}>},
{<I_{ph}^{20},$\prec^{\circ}$,I_{ph}^{23}>, <I_{ma}^{20},$\prec^{\circ}$,I_{ma}^{23}>, <I_{pl}^{20},$\prec^{\circ}$,I_{pl}^{23}>},
23>

<21,
{<I_{ph}^{21}, <PHONATION:voiceless>>, <I_{ma}^{21}, <MANNER:plosive>>, <I_{pl}^{21}, <PLACE: velar>>},
{<I_{ph}^{21},$\circ$,I_{ma}^{21}>, <I_{ma}^{21},$\circ$,I_{pl}^{21}>, <I_{ph}^{21},$\circ$,I_{pl}^{21}>},
{<I_{ph}^{21},$\prec^{\circ}$,I_{ph}^{26}>, <I_{ma}^{21},$\prec^{\circ}$,I_{ma}^{26}>, <I_{pl}^{21},$\prec^{\circ}$,I_{pl}^{26}>},
26>

<22,
{<I_{ph}^{22}, <PHONATION:voiceless>>, <I_{ma}^{22}, <MANNER:fricative>>, <I_{pl}^{22}, <PLACE: apical>>},
{<I_{ph}^{22},$\circ$,I_{ma}^{22}>, <I_{ma}^{22},$\circ$,I_{pl}^{22}>, <I_{ph}^{22},$\circ$,I_{pl}^{22}>},
{<I_{ph}^{22},$\prec^{\circ}$,I_{ph}^{28}>, <I_{ma}^{22},$\prec^{\circ}$,I_{ma}^{28}>, <I_{pl}^{22},$\prec^{\circ}$,I_{pl}^{28}>},
28>

<22,
{<I_{ph}^{22}, <PHONATION:voiceless>>, <I_{ma}^{22}, <MANNER:plosive>>, <I_{pl}^{22}, <PLACE: apical>>},
{<I_{ph}^{22},$\circ$,I_{ma}^{22}>, <I_{ma}^{22},$\circ$,I_{pl}^{22}>, <I_{ph}^{22},$\circ$,I_{pl}^{22}>},
{<I_{ph}^{22},$\prec^{\circ}$,I_{ph}^{29}>, <I_{ma}^{22},$\prec^{\circ}$,I_{ma}^{29}>, <I_{pl}^{22},$\prec^{\circ}$,I_{pl}^{29}>},
29>

<22,
{<I_{ph}^{22}, <PHONATION:voiceless>>, <I_{pl}^{22}, <PLACE: apical>>, <I_{ma}^{22}, <MANNER:α>>}
| $\alpha \in$ {plosive, fricative},
{<I_{ph}^{22},$\circ$,I_{ma}^{22}>, <I_{ma}^{22},$\circ$,I_{pl}^{22}>, <I_{ph}^{22},$\circ$,I_{pl}^{22}>},
{},
end>

<22,
{},
{},
{},
end>

<23,
{<I_{ph}^{23}, <PHONATION:voiceless>>, <I_{ma}^{23}, <MANNER:fricative>>, <I_{pl}^{23}, <PLACE: apical>>},
{<I_{ph}^{23},$\circ$,I_{ma}^{23}>, <I_{ma}^{23},$\circ$,I_{pl}^{23}>, <I_{ph}^{23},$\circ$,I_{pl}^{23}>},
{<I_{ph}^{23},$\prec^{\circ}$,I_{ph}^{25}>, <I_{ma}^{23},$\prec^{\circ}$,I_{ma}^{25}>, <I_{pl}^{23},$\prec^{\circ}$,I_{pl}^{25}>},
25>

<24,
{<I_{ph}^{24}, <PHONATION:voiced>>, <I_{ma}^{24}, <MANNER:nasal>>, <I_{pl}^{24}, <PLACE: apical>>>>},
{<I_{ph}^{25},$\circ$,I_{ma}^{24}>, <I_{ma}^{24},$\circ$,I_{pl}^{24}>, <I_{ph}^{24},$\circ$,I_{pl}^{24}>},
{},
end>

<24
{<$I_{ph}{}^{24}$, <PHONATION:voiceless>>, <$I_{ma}{}^{24}$, <MANNER:fricative>>, <$I_{pl}{}^{24}$, <PLACE: apical>>},
{<$I_{ph}{}^{24}$,$^{\circ}$,$I_{ma}{}^{24}$>, <$I_{ma}{}^{24}$,$^{\circ}$,$I_{pl}{}^{24}$>, <$I_{ph}{}^{24}$,$^{\circ}$,$I_{pl}{}^{24}$>},
{},
end>

<24,
{},
{},
{},
end>

<25,
{<$I_{ph}{}^{25}$, <PHONATION:voiceless>>, <$I_{ma}{}^{25}$, <MANNER:plosive>>, <$I_{pl}{}^{25}$, <PLACE: apical>>},
{<$I_{ph}{}^{25}$,$^{\circ}$,$I_{ma}{}^{25}$>, <$I_{ma}{}^{25}$,$^{\circ}$,$I_{pl}{}^{25}$>, <$I_{ph}{}^{25}$,$^{\circ}$,$I_{pl}{}^{25}$>},
{<$I_{ph}{}^{25}$,$\prec^{\circ}$,$I_{ph}{}^{27}$>, <$I_{ma}{}^{25}$,$\prec^{\circ}$,$I_{ma}{}^{27}$>, <$I_{pl}{}^{25}$,$\prec^{\circ}$,$I_{pl}{}^{27}$>},
27>

<26,
{<$I_{ph}{}^{26}$, <PHONATION:voiceless>>, <$I_{ma}{}^{26}$, <MANNER:fricative>>, <$I_{pl}{}^{26}$, <PLACE: apical>>},
{<$I_{ph}{}^{26}$,$^{\circ}$,$I_{ma}{}^{26}$>, <$I_{ma}{}^{26}$,$^{\circ}$,$I_{pl}{}^{26}$>, <$I_{ph}{}^{26}$,$^{\circ}$,$I_{pl}{}^{26}$>},
{},
end>

<26,
{},
{},
{},
end>

<27,
{<$I_{ph}{}^{27}$, <PHONATION:voiceless>>, <$I_{ma}{}^{27}$, <MANNER:fricative>>, <$I_{pl}{}^{27}$, <PLACE: apical>>},
{<$I_{ph}{}^{27}$,$^{\circ}$,$I_{ma}{}^{27}$>, <$I_{ma}{}^{27}$,$^{\circ}$,$I_{pl}{}^{27}$>, <$I_{ph}{}^{27}$,$^{\circ}$,$I_{pl}{}^{27}$>},
{},
end>

<28,
{<$I_{ph}{}^{28}$, <PHONATION:voiceless>>, <$I_{ma}{}^{28}$, <MANNER:fricative>>, <$I_{pl}{}^{28}$, <PLACE: apical>>},
{<$I_{ph}{}^{28}$,$^{\circ}$,$I_{ma}{}^{28}$>, <$I_{ma}{}^{28}$,$^{\circ}$,$I_{pl}{}^{28}$>, <$I_{ph}{}^{28}$,$^{\circ}$,$I_{pl}{}^{28}$>},
{},
end>

<29,
{<$I_{ph}{}^{29}$, <PHONATION:voiceless>>, <$I_{ma}{}^{29}$, <MANNER:plosive>>, <$I_{pl}{}^{29}$, <PLACE: apical>>},
{<$I_{ph}{}^{29}$,$^{\circ}$,$I_{ma}{}^{29}$>, <$I_{ma}{}^{29}$,$^{\circ}$,$I_{pl}{}^{29}$>, <$I_{ph}{}^{29}$,$^{\circ}$,$I_{pl}{}^{29}$>},
{},
end>

BIBLIOGRAPHY

[1] Aho, A. V. & J. D. Ullman (1972): *The Theory of Parsing, Translation and Compiling*. Volume 1: Parsing, Englewood Cliffs, N.J.: Prentice-Hall-Inc.

[2] Aho, A. V.; J. E. Hopcroft & J. D. Ullman (1983): *Data Structures and Algorithms*. Reading, Mass., Addison-Wesley Publishing Company.

[3] Allen, J. F. (1983): Maintaining Knowledge about Temporal Intervals. In: *Communications of the ACM 26, 11*: 832-843.

[4] Allen, J. F. & P. J. Hayes (1987): *Moments and Points in an Interval-Based Temporal Logic*. TR 180, Department of Computer Science, University of Rochester.

[5] Allerhand, M. (1987): *Knowledge-Based Speech Pattern Recognition*. London: Kogan Page.

[6] Althoff, F.; J. Carson-Berndsen; G. Drexel; D. Gibbon; K. Hübener & M. Pampel (1994): *Linguistic Word Recognition: BELLEx3 Manual*. Verbmobil Technisches Dokument 12, University of Bielefeld and University of Hamburg.

[7] Althoff, F.; J. Carson-Berndsen; G. Drexel; D. Gibbon; K. Hübener; U. Jost; K. Kirchhoff; M. Pampel; A. Petzold & V, Strom (1995): *BELLEx3+1: Linguistische Worterkennung unter Berücksichtigung der Prosodie*. Verbmobil Technisches Dokument 22, University of Bielefeld, University of Bonn and University of Hamburg.

[8] Archangeli, D. (1988): Aspects of underspecification theory. *Phonology 5*:183-207.

[9] Barton, G. E; R. C. Berwick & E. S. Ristad (1987): *Computational Complexity and Natural Language*, Cambridge, Mass. London: The MIT Press.

[10] Batóg, T. (1967): *The Axiomatic Method in Phonology*. London: Routledge & Kegan Paul Ltd.

[11] Bear, J. (1988): Morphology with Two-Level Rules and Negative Rule Features. In: *Proceedings of the 12th International Conference on Computational Linguistics (COLING-88)*, Budapest: 28-31.

[12] Bear, J. (1990): Backwards Phonology. In: *Proceedings of the 13th International Conference on Computational Linguistics (COLING-90)*, Helsinki 3: 13-20.

[13] Benthem, J. F. A. K van (1983): *The Logic of Time*. Dordrecht, Boston, London: D. Reidel Publishing Company.

[14] Berwick, R. C. & A. S. Weinberg (1982): Parsing Efficiency, Computational Complexity, and the Evaluation of Grammatical Theories. In : *Linguistic Inquiry* 13, Number 2:165-191.

[15] Berwick, R. C. & S. Pilato (1987): Learning Syntax by Automata Induction. In: *Machine Learning*, Volume 2, No. 1: 9-38.

[16] Bird, S. & E. Klein (1990): Phonological Events, In: *Journal of Linguistics* 26: 33-56.

[17] Bird S. (1990): *Constraint-Based Phonology*. PhD. Thesis, University of Edinburgh.

[18] Bird S. (1995): *Computational Phonology : A constraint-based approach*.Cambridge: Cambridge University Press.

[19] Bird, S. (ed.) (1991): *Declarative Perspectives on Phonology*, Working Papers in Cognitive Science Vol 7. University of Edinburgh.

[20] Bird, S. & D. R. Ladd (1991): Review Article: Presenting autosegmental phonology. In *Journal of Linguistics 27*: 193-210.

[21] Bird, S. & T. M. Ellison (1992): *One Level Phonology: Autosegmental Representations and Rules as Finite-State Automata*. EUCCS/RP-51, University of Edinburgh, April 1992.

[22] Bleiching, D. (1991): *Default-Hierarchien in der deutschen Wortbetonung*. ASL-TR-1991/UBI, University of Bielefeld, December 1991.

[23] Bleiching, D. (1992): Prosodisches Wissen im Lexikon. In: Görz, G: *KONVENS 92*, 1. Konferenz "Verarbeitung natürlicher Sprache", Berlin: Springer-Verlag: 59-68.

[24] Braun, G. (1991a): *SAMLAB*, Ms. University of Bielefeld.

[25] Braun, G. (1991b): Tools in Speech Technology: Problems in Segmental Labelling. Paper held at the Workshop on Computational (Morpho)Phonology, ZiF, University of Bielefeld, 23rd-25th October 1991.

[26] Broe, M. (1991): A Unification-Based Approach to Prosodic Analysis. In: Bird, S. (ed.): *Declarative Perspectives on Phonology*, Working Papers in Cognitive Science Vol 7. University of Edinburgh: 27-44.

[27] Bromberger, S. & M. Halle (1989): Why Phonology Is Different. In: *Linguistic Inquiry* 20, Number 1: 51-70.

[28] Browman, C. P. & L. Goldstein (1986): Towards an articulatory phonology. In: *Phonology Yearbook* 3: 219-252.

[29] Browman, C. P. & L. Goldstein (1989): Articulatory gestures as phonological units. In: *Phonology* 6, Cambridge: Cambridge University Press, 201-251.

[30] Browman, C. P. & L. Goldstein (1990): Tiers in articulatory phonology, with some implications for casual speech. In: Kingston, J: & M. E. Beckman (eds.): *Papers in Laboratory Phonology I: Between the Grammar and Physics of Speech*, Cambridge, Cambridge University Press: 341-376.

[31] Cahill, L. (1990): *Syllable-based Morphology for Natural Language Processing.* (DPhil Dissertation), Technical Report. Cognitive Science Research Report 181, Cognitive and Computing Sciences, University of Sussex.

[32] Cahill, L. (1993): Morphonology in the Lexicon. In: Proceedings of the 6th Conference of the European Chapter of the Association for Computational Linguistics. Utrecht: 87-96.

[33] Calder, J. & S. Bird (1991): Defaults in Underspecification Phonology. In: Bird, S. (ed.) *Declarative Perspectives on Phonology*, Working Papers in Cognitive Science Vol 7. University of Edinburgh: 107-125

[34] Carson, J. (1988): Unification and Transduction in Computational Phonology. In: *Proceedings of the 12th International Conference on Computational Linguistics (COLING-88)*, Budapest: 106-111.

[35] Carson-Berndsen, J.; D. Gibbon; K. Knäpel (1989): Interim Report 31.03.89 and Final Report 30.09.89 *Forschungsprojekt: Entwicklung phonologischer Regelsysteme und Untersuchungen zur Automatisierung der Regelerstellung für Zwecke der automatischen Spracherkennung*. Research Project financed by the Deutsche Bundespost, Ms. University of Bielefeld.

[36] Carson-Berndsen, J. (1990): Phonological Processing of Speech Variants. In: *Proceedings of the 13th International Conference on Computational Linguistics (COLING-90)*, Helsinki 3: 21-24.

[37] Carson-Berndsen, J. (1991): *Ereignisstrukturen für phonologisches Parsen*. ASL-TR-991/UBI, University of Bielefeld, August 1991

[38] Carson-Berndsen, J. (1992a): *An event-based phonotactics for German.* ASL-TR-2992/UBI, University of Bielefeld, February 1992.

[39] Carson-Berndsen, J. (1992b): *Top-down Hypothesen an der AP/LWM-Schnittstelle*. ASLMemo-46-92/UBI, University of Bielefeld, April 1992.

[40] Carson-Berndsen, J. (1992c): Computational Tools for the Development of Event Phonologies. In: Görz, G: *KONVENS 92*, 1. Konferenz "Verarbeitung natürlicher Sprache", Berlin: Springer-Verlag: 69-78.

[41] Carson-Berndsen, J. (1992d): *Tools for the development of event phonologies*. ASL-TR35-92/UBI, University of Bielefeld, December 1992.

[42] Carson-Berndsen, J. & D. Gibbon (1992): Event Relations at the Phonetics/Phonology Interface. In: *Proceedings of the 15th International Conference on Computational Linguistics (COLING-92)*, Nantes: 1269-1273.

[43] Carson-Berndsen, J.; G. Drexel (1996): Parametrised Phonological Event Parsing. In: Gibbon, D. (ed.): *Natural Language Processing and Speech Technology*. Berlin, New York: Mouton de Gruyter: 64-70.

[44] Carson-Berndsen, J.; M. Pampel (1994): *Diagnostic Evaluation in Linguistic Word Recognition*, Verbmobil Technical Report No. 38, University of Bielefeld.

[45] Chien, L-F.; K.J. Chen & L-S. Lee (1990): An augmented chart data structure with efficient word lattice parsing scheme in speech recognition applications. In: *Proceedings of the 13th International Conference on Computational Linguistics (COLING-90)*, Helsinki 2: 60-65.

[46] Chomsky, N. (1964): *Current Issues in Linguistic Theory*. The Hague: Mouton.

[47] Chomsky, N. & M. Halle (1968): *The Sound Pattern of English*. New York: Harper and Row.

[48] Church, K. W. (1983): *Phrase Structure Parsing. A method for taking advantage of allophonic constraints*. Ph.D. Thesis, MIT. Published Indiana University Linguistics Club.

[49] Church, K. W. (1987a): Phonological parsing and lexical retrieval. In: U. H. Frauenfelder & L. Komisarjevsky Tyler (eds.): *Spoken Word Recognition* Amsterdam: Elsevier Science Publishers B.V.: 53-70.

[50] Church, K. W. (1987b): *Phonological parsing in speech recognition*. Boston: Kluwer Academic Publishers.

[51] Clements, G. N. & S. J. Keyser (1983): *CV Phonology: A Generative Theory of the Syllable*. Cambridge, Mass. London: The MIT Press.

[52] Coleman, J. (1990): Unification Phonology: Another look at "synthesis-by-rule". In: *Proceedings of the 13th International Conference on Computational Linguistics (COLING-90)*, Helsinki 3: 79-84.

[53] Coleman, J. (1992): The phonetic interpretation of headed phonological structures containing overlapping constituents. In: *Phonology 9*: 1-44.

[54] Craig, I. D. (1989): *The CASSANDRA Architecture: distributed control in a blackboard system*. Chichester: Ellis Horwood Limited.

[55] Davis, S. (1992): Investigating English phonotactic constraints using a computerized lexicon. In: Klenk, U. (ed.): *Computatio Linguae*, Stuttgart: Franz Steiner Verlag: 1-15.

[56] Dogil, G & H. C. Luschützky (1990): Notes on sonority and segmental strength. In *Rivista di Linguistica 2,2*: 3-52.

[57] Douglas, S. & R. Dale (1992): Towards Robust PATR. In: *Proceedings of the 15th International Conference on Computational Linguistics (COLING-92)*, Nantes: 468-474.

[58] *Duden Aussprachewörterbuch*: Wörterbuch der deutschen Standardaussprache. Bearbeitet von Max Mangold et al. Aufl. 1962., völlig neu bearbeitete und erweiterte Aufl. 1974. Bibliographisches Institut Mannheim.

[59] Earley, J. (1970): An efficient context-free parsing algorithm. In: *Communications of the Association for Computing Machinery* 13: 94-102.

[60] Eikmeyer, H-J.; W. Kindt; U. Laubenstein; S. Lisken; T. Polzin; H. Rieser & U. Schade (1991): Kohärenzkonstitution im gesprochenen Deutsch. In: Rickheit, G. (ed.): *Kohärenzprozesse*. Opladen: Westdeutscher-Verlag.

[61] Erman, L. D. & V. R. Lesser (1990): The HEARSAY-II Speech Understanding System: A Tutorial. In: Waibel, A & K-F. Lee (1990): *Readings in Speech Recognition*. San Mateo, California: Morgan Kaufmann Publishers, Inc.: 235-245.

[62] Fant, G. (1973): *Speech Sounds and Features*. Cambridge, Mass.: MIT Press.

[63] Faust, L. & B. Möbius (1992): *Kriterien für das Etikettieren von Ereignissen*. ASL-TR55-92/UBN, University of Bonn, August, 1992.

[64] Fourcin, A. J.; G. Harland; W. Barry & V. Hazan (eds.)(1989): *Speech Input and Output Assessment: Multilingual Methods and Standards*. Chichester: Elis Horwood.

[65] Frauenfelder, U. H. & A. Lahiri (1989): Understanding Words and Word Recognition: Can Phonology Help? In: Marslen-Wilson, W. D. (ed.): *Lexical Representation and Process*. Cambridge, Mass., London: The MIT Press.

[66] Freksa, C.(1992): Temporal reasoning based on semi-intervals. In: *Artificial Intelligence 54*, Elsevier Science Publishers B.V.:199-227.

[67] Fujimura, O. (1976): Syllables as concatenated demisyllables and affixes. In: *Journal of the Acoustical Society of America* 59(1): S55.

[68] Gazdar, G. (1985): Review article:Finite state morphology. In: *Linguistics 23*: 597-607.

[69] Gibbon, D. (1985): Prosodic Parsing in English and Akan. Paper given at the *21st International Conference on Contrastive Linguistics*, Blazejewko, Poland.

[70] Gibbon, D. (1987): Finite State Processing of Tone Systems. In: *Proceedings of the Third Conference of the European Chapter of the Association for Computational Linguistics*. Copenhagen: 291-297.

[71] Gibbon, D. (1990a): Prosodic Association by Template Inheritance. In: Gazdar, G. & W. Daelemans (eds.): *Inheritance in Natural Language Processing: Workshop Proceedings*. Tilburg: ITK: 65-81.

[72] Gibbon, D. (1990b): Linguistische Wortmodellierung: Phonologie und Morphologie. In: von Hahn, W. et al. : *Program-Architekturen von Systemen zur intergrierten Anaylse von Sprachlauten und Sprachstrukturen*. ITM 8901 3, University of Hamburg, Chapter 4.4.

[73] Gibbon, D. (1992a): *ILEX: A linguistic approach to computational lexica*. In: Klenk, U. (ed.): *Computatio Linguae*, Stuttgart: Franz Steiner Verlag: 32-53.

[74] Gibbon, D. (1992b): Prosody, Time Types and Linguistic Design Factors in Spoken Language System Architectures. In: Görz, G: *KONVENS 92*, 1. Konferenz "Verarbeitung natürlicher Sprache", Berlin: Springer-Verlag: 90-99.

[75] Gibbon, D. (1992c): *Lecture Notes on Computational Lexicology*, Fourth European Summer School in Logic, Language and Information, Essex.

[76] Gibbon, D. & G. Braun (1998a): The PSI/PHI architecture for prosodic parsing. In: *Proceedings of the 12th International Conference on Computational Linguistics (COLING-88)*, Budapest: 202-204.

[77] Gibbon, D. & G. Braun (1988b): Phonologische Netze. In Kindermann, J. & C. Lischka (eds.) *Workshop Konnektionismus*: 157-177.

[78] Gibbon, D & H. Langer (1992): Linguistische Wortmodellierung. In Reimann, D. (ed.): *Beiträge des ASL-Lexikonworkshops*, ASL-TR-40-92/ZSB: 50-56.

[79] Gibbon, D.; D. Bleiching; J. Carson-Berndsen, H. Langer & M. Pampel (1992): *BELLEx3: Bielefeld Engine for Lattice-to-Lattice Event-parsing*. Ms. University of Bielefeld, December 1992.

[80] Gibbon, D.; R. Moore & R. Winski (eds.) (1997): *Handbook of Standards and Resources for Spoken Language Systems.* Berlin, New York: Mouton de Gruyter.

[81] Giegerich, R. & D. Wheeler (1996): Lecture Notes on Pairwise Alignment. World-Wide-Web Document URL: http://merlin.mbcr.bcm.tmc.edu:8001/bcd/Curric/PrwAli/prwali.html

[82] Gimson, A. C. (1970): *An Introduction to the Pronunciation of English.* 2nd Edition. London: Edward Arnold.

[83] Görz, G. (1988): *Strukturanalyse natürlicher Sprache.* Bonn: Addison-Wesley Publishing Company.

[84] Görz, G. (1992): *Kognitiv orientierte Architekturen für die Sprachverarbeitung.* ASLTR-39-92/UER, University of Erlangen-Nürnberg, September 1992.

[85] Görz, G. & D. Paulus (1988): A finite state approach to German verb morphology. In: *Proceedings of the 12th International Conference on Computational Linguistics (COLING-88),* Budapest: 212-215.

[86] Görz, G. & H. Weber (1992): *Incremental Processing in ASL.* ASL-TR-38-92/UER, University of Erlangen-Nürnberg, March 1992.

[87] Goldsmith, J. (1976): *Autosegmental Phonology.* Bloomington, Indiana: Indiana University Linguistics Club.

[88] Goldsmith, J. (1990): *Autosegmental and Metrical Phonology.* Cambridge, Mass.: Basil Blackwell Inc.

[89] Grosjean, F & J. P. Gee (1987): Prosodic structure and spoken word recognition. In: U. H. Frauenfelder & L. Komisarjevsky Tyler, (eds.): *Spoken Word Recognition* Amsterdam: Elsevier Science Publishers B.V.: 135-156.

[90] Hahn, W. von (1992): Von der Verknüpfung zur Integration: Kontrollstrategie oder kognitive Architektur. In: Görz, G: *KONVENS 92*, 1. Konferenz "Verarbeitung natürlicher Sprache", Berlin: Springer-Verlag: 1-10.

[91] Halle, M. (1959): *The sound pattern of Russian.* The Hague: Mouton & Co.

[92] Halle, M. (1964): On the Bases of Phonology. In : Fodor, J. A. & J. J. Katz (eds.): *The Structure of Language: Readings in the Philosophy of Language.* Englewood Cliffs: Prentice-Hall Inc.: 324-333.

[93] arris, Z. S. (1961): *Structural Linguistics.* Chicago: Phoenix Books.

[94] 't Hart, J. & R. Collier (1975): Integrating different levels of intonation analysis. In: *Journal of Phonetics* 3: 235-255.

[95] Hauenstein, A. & K. Hübener (1992): Towards Cognitive Processing of Speech Signals. In: Görz, G: *KONVENS 92*, 1. Konferenz "Verarbeitung natürlicher Sprache", Berlin: Springer-Verlag: 100-108.

[96] Hirschman, L; & H. Thompson (1994): *Overview of evaluation in speech and natural language processing.* Unpublished Draft, 1994.

[97] Hockett, C. F. (1955): *A Manual of Phonology.* Baltimore: Waverly Press.

[98] Holmes, J. N. (1988): *Speech Synthesis and Recognition.* Wokingham: Van Nostrand Reinhold (UK) Co. Ltd.

[99] Hooper, J. (1976): *Introduction to Natural Generative Phonology.* New York: Academic Press

[100] Hopcroft, J. E. & J. D. Ullman (1979): *Introduction to Automata Theory, Languages and Computation.* Reading, Mass. : Addison-Wesley Publishing Company.

[101] Hübener, K. (1993): *Detektion akustisch-phonetischer Ereignisse.* Verbmobil Memo 5, University of Hamburg.

[102] übener, K. & J. Carson-Berndsen (1994): Phoneme recognition using acoustic events. In *Proceedings of the 3rd International Conference on Spoken Languauge Processing, vol 4*, Yokohama, Japan, 1994: 1919 - 1922.

[103] Hulst, H van der & N. Smith (1982): *The structure of phonological representations* (Part I), Dordrecht: Foris Publications.

[104] Jespersen, O. (1904): *Lehrbuch der Phonetik.* Leipzig: B. G. Teubner.

[105] Johnson, C. D. (1972): *Formal Aspects of Phonological Description.* The Hague: Mouton & Co. N.V, Publishers.

[106] Jusek,A.; H. Rautenstrauch; G.A. Fink; F. Kummert; G. Sagerer; J. Carson-Berndsen & D. Gibbon (1994): Dektektion unbekannter Wörter mit Hilfe phonotaktischer Modelle. In *Mustererkennung 94, 16. DAGM-Symposium Wien*, Berlin, 1994. Springer–Verlag.

[107] Joos, M. (1957): *Readings in Linguistics I.* (4th Edition). Chicago, London: The University of Chicago Press.

[108] Kahn, D. (1976): *Syllable-Based Generalisations in English Phonology* Indiana: Indiana University Linguistics Club.

[109] Kaplan, R. M & M. Kay (1981): Phonological rules and finite-state transducers. Paper held at the Annual Meeting of L.S.A in New York.

[110] Kaplan, R. M. & M. Kay (1994): Regular models of phonological rule systems. In *Computational Linguistics 20*: 331-378

[111] Karttunen, L. & K. Wittenburg (1983): A two-level morphological analysis of English. *Texas Linguistic Forum 22*: 217-228.

[112] Karttunen, L. (1991): *Rules and Transducers*. International Conference on Current Issues in Computational Linguistics, Universiti Sains Malaysia, Penang, Malaysia.

[113] Kasper, R. T. & W. C. Rounds (1986): A Logical Semantics for Feature Structures. In: Proceedings of the 24th Annual Meeting of the Association for Computational Linguistics. New York: 257-266.

[114] Katz, J. J. & J. A. Fodor (1964): The Structure of a Semantic Theory. In : Fodor, J. A. & J. J. Katz (eds.): *The Structure of Language: Readings in the Philosophy of Language*. Englewood Cliffs: Prentice-Hall Inc.:479-518.

[115] Kay, M. (1983): When meta-rules are not meta-rules. In: Sparck-Jones, K. & Y. A. Wilks eds.: *Automatic Natural Language Parsing*. Chichester/New York: Ellis Horwood/Wiley: 84-116.

[116] Kay, M. (1987): Nonconcatenative Finite-State Morphology. In: *Proceedings of the Third Conference of the European Chapter of the Association for Computational Linguistics*. Copenhagen: 2-10.

[117] Kaye, J. (1989): *Phonology: a cognitive view*. Hillsdale, New Jersey: Lawrence Erlbaum Associates, Publishers.

[118] Keating, P.A. (1988): Underspecification in phonetics. *Phonology 5*:275-292.

[119] Kemke, C. (1990): *Konnektionistische Modelle im Bereich Speech and Language*. In: von Hahn, W. et al. : *Program-Architekturen von Systemen zur intergrierten Anaylse von Sprachlauten und Sprachstrukturen*. ITM 8901 3, University of Hamburg, Chapter 3.5.

[120] Kenstowicz, M & C. Kisseberth (1979): *Generative Phonology*. New York, San Francisco, London: Academic Press.

[121] Kilbury, J. (1984): *Earley-basierte Algorithmen für direktes Parsen mit ID/LPGrammatiken*. KIT-Report 16, Technical University Berlin.

[122] Kilbury, J. (1985): Chart Parsing and the Earley Algorithm. In: Klenk, U. (ed.): *Kontextfreie Syntaxen und Verwandte Systeme*. Tübingen: Max Niemeyer Verlag

[123] Kimball, J. (1973): Seven Principles of Surface Structure Parsing in Natural Language. In: *Cognition* 2/1: 15-47.

[124] Kiparsky, P. (1982): From Cyclic Phonology to Lexical Phonology. In: Hulst, H van der & N. Smith (eds.): *The structure of phonological representations* (Part I), Dordrecht: Foris Publications: 131-176.

[125] Kirchhoff, K. (1995a): *Two-level Modelling of Speech Variant Rules.* Verbmobil Technical Report No. 82. University of Bielefeld.

[126] Kirchhoff, K. (1995b): *Phonetic Features in Speech Recognition: A Delayed Synchronisation Approach.* Master's Thesis, University of Bielefeld.

[127] Kirchhoff, K. (1996): Phonologisch strukturierte HMMs. In: Gibbon, D. (ed.): *Natural Language Processing and Speech Technology.* Berlin, New York: Mouton de Gruyter: 55-63.

[128] Klatt, D. H. (1979): Speech perception: a model of acoustic-phonetic analysis and lexical access. In: *Journal of Phonetics* 7: 279-312.

[129] Kloeke, W. van Lessen (1982): *Deutsche Phonologie und Morphologie: Merkmale und Markiertheit.* Tübingen: Max Niemeyer Verlag.

[130] Kohler, K. J. (1977): *Einführung in die Phonetik des Deutschen.* Berlin: Erich Schmidt Verlag.

[131] ornai, A. (1991): *Formal Phonology*, Ph.D. Thesis, Stanford University.

[132] Koskenniemi, K. (1983a): Two-level model for morphological analysis. In: *Proceedings of the Eighth International Joint Conference on Artificial Intelligence, (IJCAI83)*: 683-685.

[133] Koskenniemi, K. (1983b): *Two-Level Morphology: A general computational model for word-form recognition and production.* University of Helsinki, Department of General Linguistics, Publications No. 11.

[134] Koskenniemi, K. & K. W. Church (1988): Complexity, Two-Level Morphology and Finnish. In: *Proceedings of the 12th International Conference on Computational Linguistics (COLING-88),* Budapest: 335-339.

[135] Krech, H.; E.-M. Krech & H. Stelzig (eds.) (1969): *Wörterbuchder deutschen Aussprache*, 2nd Edition, Munich: Max Heuber Verlag.

[136] Kunzmann, S. (1991): *Die Worterkennung in einem Dialogsystem für kontinuierlich gesprochene Sprache.* Tübingen: Max Niemeyer Verlag.

[137] Ladefoged, P. (1990): On dividing phonetics and phonology: comments on the papers by Clements and by Browman & Goldstein. In: Kingston, J: & M. E. Beckman (eds.): *Papers in Laboratory Phonology I: Between the Grammar and Physics of Speech*, Cambridge, Cambridge University Press: 398-405.

[138] Lancé, D,; W. Hess; L. Faust; B. Möbius & M. Pätzold (1991): *Das Ereigniskonzept unter akustisch-phonetischem Aspekt*, ASL-TR-16-91/UBN, University of Bonn, November 1991.

[139] Lakoff, G. (1987): A Suggestion For a Linguistics With Connectionist Foundations. In: Touretzky, D. et al. (eds): *Proceedings of the 1987 Connectionist Models Summer School*, Carnegie-Mellon University: 301-314.

[140] Langer, H. (1990): Syntactic Normalisation of Spontaneous Speech. In: *Proceedings of the 13th International Conference on Computational Linguistics (COLING-90)*, Helsinki 3: 180-183.

[141] Langer, H. (1991): *Grundbegriffe der ASL-Architektur*. ASL-Memo-13-91/UBI, University of Bielefeld, October 1991.

[142] Langer, H. & D. Gibbon (1992): *DATR as a graph representation language for ILEX speech oriented lexica*. ASL-TR-43-92/UBI, University of Bielefeld, March 1992.

[143] Lehning, M. (1994): *Ein Programmsystem zur Evaluierung der signalnahen Spracherkennung*. Verbmobil Technical Report No. 9, Technical University Braunschweig.

[144] Liberman,M. & A. Prince (1977): On Stress and Linguistic Rhythm. In: *Linguistic Inquiry* 8: 249-336.

[145] Lippmann, R. P. (1989): Review of Neural Networks for Speech Recognition. In: Waibel, A & K-F. Lee (1990): *Readings in Speech Recognition*. San Mateo, California: Morgan Kaufmann Publishers, Inc.: 374-392

[146] Marslen-Wilson, W. D. (1987): Functional parallelism in spoken word-recognition. In: U. H. Frauenfelder & L. Komisarjevsky Tyler, (eds.): *Spoken Word Recognition* Amsterdam: Elsevier Science Publishers B.V.: 71-102.

[147] McClelland, J. L.; J. L. Elman (1986): The TRACE model of speech perception. In: *Cognitive Psychology* 18: 1-86.

[148] Niemann, H; M. Lang & G. Sagerer (1988): *Recent Advances in Speech Understanding and Dialog Systems*. Berlin, Heidelberg: Springer-Verlag.

[149] Pampel, M. (1992): *Die Repräsentation lexikalischen phonologischen Wissens am Beispiel der Wortbetonung*. ASL-TR-58-92/UBI, University of Bielefeld, September 1992.

[150] Partee, B. H.; A. ter Meulen & R. E. Wall (1990): *Mathematical Methods in Linguistics*. Dordrecht, Boston, London: Kluwer Academic Publishers.

[151] Philipp, M. (1974): *Phonologie des Deutschen*. Stuttgart, Berlin, Köln, Mainz: Verlag W. Kohlhammer.

[152] Pierrehumbert, J. B. (1980): *The phonology and phonetics of English intonation*. Cambridge, Mass.: MIT Diss.

[153] Pike, K. & E. Pike (1947): Immediate constituents of Mazatec Syllables. In: *International Journal of American Linguistics* 13: 78-91.

[154] Pollard, C. & I. A. Sag (1987): *Information-Based Syntax and Semantics* Volume 1, Fundamentals, CSLI, Lecture Notes Number 13.

[155] rince, A & P. Smolemsky (1993): *Optimality Theory: Constraint Interaction in Generative Grammar.* TR-2, Center for Cognitive Science, Rutgers University.

[156] Pyka, C. (1991): *Deterministische, inkrementelle und zeitsynchrone Verarbeitung und die Architektur von ASL-Nord.* ASL-TR-22-91/UHH, University of Hamburg, December 1991.

[157] Rabiner, L. R. (1989): A Tutorial on Hidden Markov Models and Selected Applications in Speech Recognition. In: Waibel, A & K-F. Lee (1990): *Readings in Speech Recognition.* San Mateo, California: Morgan Kaufmann Publishers, Inc.: 267-296.

[158] Reape, M. & H. Thompson (1988): Parallel intersection and serial composition of finite state transducers. In: *Proceedings of the 12th International Conference on Computational Linguistics (COLING-88),* Budapest: 535-539.

[159] Reinhard, S. (1990): *Adäquatheitsprobleme automatenbasierter Morphologiemodelle am Beispiel der deutschen Umlautung.* Masters Thesis, Ms. University of Trier.

[160] Reinhard, S. & D. Gibbon (1991): Prosodic Inheritance and Morphological Generalisations. In: Proceedings of the Fifth Conference of the European Chapter of the Association for Computational Linguistics. Berlin: 131-136.

[161] Ritchie, G. D.; G. J. Russell; A. W. Black & S. G. Pulman (1992): *Computational Morphology.* Cambridge, Mass. London: The MIT Press.

[162] Sagerer, G. (1990): *Automatisches Verstehen gesprochener Sprache.* Mannheim, Wien, Zürich: BI-Wissenschaftsverlag.

[163] Sagey, E. (1988): On the Ill-Formedness of Crossing Association Lines. In: *Linguistic Inquiry* 19, Number 1: 109-118

[164] Sampson, G. (1970): *Stratificational Grammar.* The Hague: Mouton & Co. N.V., Publishers

[165] Schade, U. (1987): *'Fischers Fritz fischt fische Fische' - Konnektionistische Modelle der Satzproduktion.* KoLiBri Arbeitsbericht-Nr. 6, University of Bielefeld.

[166] Schane, S. A. (1973): *Generative Phonology*. Englewood Cliffs, New Jersey: Prentice Hall, Inc.

[167] Scholz, H.-J. (1972): *Untersuchungen zur Lautstruktur deutscher Wörter*. München: Wilhelm Fink Verlag.

[168] Schotola, T. (1984): On the use of demisyllables in automatic word recognition. In: *Speech Communication* 5: 63-87.

[169] Schukat-Talamazzini, E.G.; S. Rieck & T. Kuhn (1991): *Handbuch der Automatischen Spracherkennung*, Ms. University of Erlangen-Nürnberg.

[170] Scobbie, J. (1991a): Towards Declarative Phonology. In: S. Bird (ed.): *Declarative Perspectives on Phonology*, Edinburgh Working Papers in Cognitive Science, Vol 7, Centre for Cognitive Science, The University of Edinburgh.

[171] Scobbie, J. (1991b): *Attribute value phonology*. Ph.D. Thesis, University of Edinburgh.

[172] Scobbie, J.; J. Coleman & S. Bird (1996): Key Aspects of Declarative Phonology. In: Durand, J. & B. Laks: *Current Trends in Phonology: Models and Methods*, European Studies Research Institute, University of Salford, Vol 2.: 685-710.

[173] Seck, R. & G. Ruske (1986): Structure of German syllable initial and final consonant clusters based on articulatory features. In: *Speech Communication* 5: 347-354.

[174] Selkirk, E. O. (1982): The Syllable. In: van der Hulst, H. & N. Smith eds.: *The structure of phonological representations* (Part II), Dordrecht, Holland: Foris Publications.

[175] Shannon, C.E. & W. Weaver (1949): *The Mathematical Theory of Communication*. Urbana, Chicago, London: University of Illinois Press.

[176] Shieber, S. M. (1986): *An Introduction to Unification-Based Approaches to Grammar*. Stanford: CLSI.

[177] Sommerstein, A. H. (1977): *Modern Phonology*. London: Edward Arnold.

[178] Stampe, D. & P. J. Donegan (1979): The Study of Natural Phonology. In: Dinnsen, D. A. (ed): *Current Approaches to Phonological Theory*. Bloomington, London: Indiana University Press: 126-173.

[179] Ternes, E. (1987): *Einführung in die Phonologie*. Darmstadt: Wissenschaftliche Buchgesellschaft.

[180] Treiman, R. (1986): The Division between Onsets and Rimes in English Syllables. In: *Journal of Memory and Language* 25, 476-491.

[181] Trost, H. (1990): The application of two-level morphology to non-concatenative German morphology. In: *Proceedings of the 13th International Conference on Computational Linguistics (COLING-90)*, Helsinki 2: 371-376.

[182] Vennemann, T. (1972): On The theory of syllabic phonology. In: *Linguistische Berichte* 18: 1-18.

[183] Vennemann, T. (1988): *Preference Laws for Syllable Structure: and the Explanation of Sound Change*. Berlin, New York, Amsterdam: Mouton de Gruyter.

[184] Waibel, A & K-F. Lee (1990): *Readings in Speech Recognition*. San Mateo, California: Morgan Kaufmann Publishers, Inc.

[185] Weber, H. (1991): *Inkrementelle Architektur und Lexikonzugriffe*. ASL-TR-27-91/UER, University of Erlangen-Nürnberg, November 1991.

[186] Weber, H. (1992): *Chartparsing in ASL-Nord: Berichte zu den Arbeitspaketen P1 bis P9*. ASL-TR-28-92/UER, University of Erlangen-Nürnberg, December 1992.

[187] Wells, J. C. (1989): Computer-coded Phonemic Notation of Individual Languages of the European Community. In: *Journal of the International Phonetic Association* 19(1): 31-54.

[188] Wheeler, D. W. & D. S. Touretzky (1989): *A Connectionist Implementation of Cognitive Phonology*. CMU-CS-89-144.

[189] Wheeler, D. W. & D. S. Touretzky (1990): From Syllables to Stress: A Cognitively Plausible Model. In: Touretzky, D. S; D. W. Wheeler & G. Elvgren III (1990): *Rules and Maps III: Further Progress in Connectionist Phonology*, CMU-CS-90138: 17-31.

[190] Whorf, B. L. (1940): Linguistics as an Exact Science. Reprinted 1956 in: *Language, Thought and Reality* Selected writings of Benjamin Lee Whorf, Edited by J. B. Carroll. Cambridge, Mass.: The MIT Press: 220-232.

[191] Wiebe, B. (1992): *Modelling Autosegmental Phonology with Multi-Tape Finite State Transducers*. Masters Thesis, Ms. Simon Fraser University.

[192] Wiese, R. (1990): Towards a Unification-Based Phonology. In: *Proceedings of the 13th International Conference on Computational Linguistics (COLING-90)*, Helsinki 3:283-286.

[193] Williams, S. (1990): *Lexical Phonology: A computational system*. CS-91-03, University of Sheffield.

[194] Wirén M. (1989): Interactive Incremental Chart Parsing. In: *Proceedings of the Fourth Conference of the European Chapter of the Association for Computational Linguistics*. Manchester: 241-248.

[195] Wirén M. (1990): Incremental Parsing and Reason Maintenance. In: *Proceedings of the 13th International Conference on Computational Linguistics (COLING-90)*, Helsinki 3: 287-292.

INDEX

Text, Speech and Language Technology

1. H. Bunt and M. Tomita (eds.): *Recent Advances in Parsing Technology.* 1996 ISBN 0-7923-4152-X
2. S. Young and G. Bloothooft (eds.): *Corpus-Based Methods in Language and Speech Processing.* 1997 ISBN 0-7923-4463-4
3. T. Dutoit: *An Introduction to Text-to-Speech Synthesis.* 1997 ISBN 0-7923-4498-7
4. L. Lebart, A. Salem and L. Berry (eds.): *Exploring Textual Data.* 1998 ISBN 0-7923-4840-0
5. J. Carson-Berndsen, *Time Map Phonology.* 1998 ISBN 0-7923-4883-4

KLUWER ACADEMIC PUBLISHERS – DORDRECHT / BOSTON / LONDON